CONGRESSIONAL POLICIES, PRACTICES AND PROCEDURES

KEY CONGRESSIONAL REPORTS FOR JUNE 2019

INTERNATIONAL RELATIONS

CONGRESSIONAL POLICIES, PRACTICES AND PROCEDURES

Additional books and e-books in this series can be found on Nova's website under the Series tab.

CONGRESSIONAL POLICIES, PRACTICES AND PROCEDURES

KEY CONGRESSIONAL REPORTS FOR JUNE 2019

INTERNATIONAL RELATIONS

TARA T. ROBERTSON
EDITOR

snova
New York

Library of Congress Cataloging-in-Publication Data

ISBN: 978-1-53616-666-8

Published by Nova Science Publishers, Inc. † New York

Contents

PREFACE

This book is a comprehensive compilation of all reports, testimony, correspondence and other publications issued by the Congressional Research Service on International Relations during the month of June.

Chapter 1 - Prior to the initiation of economic reforms and trade liberalization nearly 40 years ago, China maintained policies that kept the economy very poor, stagnant, centrally controlled, vastly inefficient, and relatively isolated from the global economy. Since opening up to foreign trade and investment and implementing free-market reforms in 1979, China has been among the world's fastest-growing economies, with real annual gross domestic product (GDP) growth averaging 9.5% through 2018, a pace described by the World Bank as "the fastest sustained expansion by a major economy in history." Such growth has enabled China, on average, to double its GDP every eight years and helped raise an estimated 800 million people out of poverty. China has become the world's largest economy (on a purchasing power parity basis), manufacturer, merchandise trader, and holder of foreign exchange reserves. This in turn has made China a major commercial partner of the United States. China is the largest U.S. merchandise trading partner, biggest source of imports, and third-largest U.S. export market. China is also the largest foreign holder of U.S. Treasury securities, which help fund the federal debt and keep U.S. interest rates low. As China's economy has matured, its real GDP growth has slowed significantly, from 14.2% in 2007 to 6.6% in 2018, and that growth is projected by the International Monetary Fund (IMF) to fall to 5.5% by 2024. The Chinese government has embraced slower economic growth, referring to it as the "new normal" and acknowledging the need for China to embrace a new growth model that relies less on fixed investment and exporting, and more on private consumption, services, and innovation to drive economic growth. Such reforms are needed in order for China to avoid hitting the "middle-income trap," when countries achieve a certain economic level but begin to experience sharply diminishing economic growth rates because they are unable to adopt new sources of economic growth, such as innovation. The Chinese government has made innovation a top priority in its economic planning through a number of high-profile initiatives, such as "Made in China 2025," a plan announced in 2015 to upgrade and modernize China's manufacturing in 10 key sectors through extensive government assistance in order to make China a major global player in these sectors. However, such measures have increasingly raised concerns that China intends to use industrial policies to decrease the country's reliance on foreign technology (including by locking out foreign firms in China) and eventually dominate global markets. In 2017, the Trump Administration launched a Section 301 investigation of China's innovation and intellectual property policies deemed harmful to U.S. economic interests. It subsequently raised tariffs by 25% on $250 billion worth of imports from China, while China increased tariffs (ranging from 5% to 25%) on $110 billion worth of imports from the United States. Such measures have sharply decreased bilateral trade in 2019. On May 10, 2019, President Trump announced he was considering raising tariffs on nearly all remaining products from China. A protracted and escalating trade conflict between the United States and China could have negative consequences for the Chinese economy. China's growing global economic influence and the economic and trade policies it maintains have significant implications for the United States and hence are of major interest to Congress. While China is a large and growing market for U.S. firms, its incomplete transition to a free-market economy has resulted in economic policies deemed harmful to U.S. economic interests, such as industrial policies and theft of U.S.

intellectual property. This chapter provides background on China's economic rise; describes its current economic structure; identifies the challenges China faces to maintain economic growth; and discusses the challenges, opportunities, and implications of China's economic rise for the United States.

Chapter 2 - India, a federal republic and the world's most populous democracy, held elections to seat a new lower house of parliament in April and May of 2019. Estimates suggest that more than two-thirds of the country's nearly 900 million eligible voters participated. The 545-seat Lok Sabha (People's House) is seated every five years, and the results saw a return to power of the Bharatiya Janata Party (BJP) led by Prime Minister Narendra Modi, who was chief minister of the west Indian state of Gujarat from 2001 to 2014. Modi's party won decisively—it now holds 56% of Lok Sabha seats and Modi became the first Indian leader to win consecutive majorities since Indira Gandhi in 1971. The United States and India have been pursuing an expansive strategic partnership since 2005. The Trump Administration and many in the U.S. Congress welcomed Modi's return to power for another five-year term. Successive U.S. Presidents have deemed India's growing power and influence a boon to U.S. interests in Asia and globally, not least in the context of balancing against China's increasing assertiveness. India is often called a preeminent actor in the Trump Administration's strategy for a "free and open Indo-Pacific." Yet there are potential stumbling blocks to continued development of the partnership. In 2019, differences over trade have become more prominent, and India's long-standing (and mostly commercial) ties to Russia and Iran may run afoul of U.S. sanctions laws. Additionally, India maintains a wariness of U.S. engagement with Pakistan and intentions in Afghanistan, with Islamabad presently facilitating a U.S.-Taliban dialogue and India counseling against a precipitous U.S. withdrawal from Afghanistan. Prime Minister Modi's return to power promises broad continuity, even with some notable changes to the federal cabinet. By many accounts, Modi's record as an economic reformer and liberalizer is mixed, and his reputation as a nationalist "watchman" has not always translated into effective foreign policy, according to some analysts.

It is unclear if Modi will use his renewed domestic political mandate to pursue more assertiveness internationally, possibly in ways that challenge U.S. preferences. Still, most analysts contend that Modi and the BJP have been and will continue to be more open to aligning with U.S. regional strategy and more energetic in pursuing U.S.-favored economic reforms than would have been any alternative Indian leadership. The BJP is a Hindu nationalist party, born in 1980 of a larger social movement, and Narendra Modi is a self-avowed Hindu nationalist (India is roughly 79% Hindu and 14% Muslim). The 2019 Modi-BJP campaign was widely criticized for divisiveness, and nationalist fervor following a February India-Pakistan crisis may have benefitted the BJP at the polls. India's minority communities and the country's civil society are widely reported to be under increasing threats emanating from Hindu majoritarian policies and sentiment. These threats can take violent and repressive forms, at times with the involvement of Indian officials or political figures, as reported by the U.S. State Department and independent human rights watchdogs, and as criticized by some Members of Congress. This chapter reviews the recent Indian election process and results, the country's national political stage, and possible implications for U.S. interests in the areas of bilateral economic and trade relations, defense and security ties, India's other foreign relations, and human rights concerns.

Chapter 3 - U.S. economic sanctions imposed on North Korea are instigated by that country's activities related to weapons proliferation, especially its tests since 2006 of nuclear weapons and missile technology; regional disruptions; terrorism; narcotics trafficking; undemocratic governance; and illicit activities in international markets, including money laundering, counterfeiting of goods and currency, and bulk cash smuggling. The sanctions have the following consequences for U.S.- North Korea relations:

- Trade is limited to food, medicine, and other humanitarian-related goods, all of which require a license. Imports from North Korea are prohibited as of June 2011; exports to North Korea of most U.S.-origin goods, services, or technology are prohibited as of March 2016. Trade in luxury goods is banned. The Department of Commerce denies export licenses for reasons of nuclear proliferation, missile technology, U.N. Security Council requirements, and international terrorism.

- Arms sales and arms transfers are fully denied.
- Financial transactions are prohibited. U.S. persons are prohibited from providing financial services for the purpose of evading sanctions, or from providing financial services to a person or entity designated for sanctions. The President, in September 2017, authorized the Secretary of the Treasury to designate for sanctions any foreign financial institution that conducts or facilitates "any significant transaction on behalf of any [designated] person," or "in connection with trade with North Korea."
- North Korea is designated as a jurisdiction of primary money laundering concern by the Department of the Treasury's Financial Crimes Enforcement Network (FinCEN), effective December 9, 2016.
- U.S. new investment is prohibited, and investment in North Korea's transportation, mining, energy, or financial sectors is prohibited. North Korea is also ineligible to participate in any U.S. government program that makes credit, credit guarantees, or investment guarantees available.
- U.S. foreign aid is minimal and mostly limited to refugees fleeing North Korea; broadcasting into the country; nongovernmental organization programs dedicated to democracy promotion, human rights, and governance; and emergency food aid. In past years, aid related to disabling and dismantling the country's nuclear weapons program has been made available. By law, U.S. representatives in the international financial institutions (IFI) are required to vote against any support for North Korea due to its nuclear weapons ambitions and international terrorism. Human rights and environmental activities would also likely result in U.S. objections to North Korea's participation in the IFI.

- U.S.-based assets are blocked for North Korean individuals, entities, aircraft, and vessels designated by the Department of the Treasury's Office of Foreign Assets Control (OFAC). U.S. persons are prohibited from entering into trade and transaction with these designees and, foreign financial institutions could become subject to U.S. sanctions for facilitating transactions for designated persons.
- Kim Jong-un, the Korean Workers' Party, and others—banks, shipping companies, seagoing vessels, state agencies, and other individuals affiliated with the state's security regime—are identified as being among those engaged in illicit and punishable activities, possibly including nuclear or ballistic missile programs, undermining cybersecurity, censorship, and sanctions evasion. As a result, effective March 15, 2016, any of their assets under U.S. jurisdiction are frozen, and U.S. persons and entities are prohibited from entering into trade and transactions with the designees.
- U.S. travel requires a special validation passport issued by the State Department. Such passports are reserved only for travel in the U.S. national interest and are intended for professional reporters, officials with the American Red Cross or International Committee of the Red Cross, or those who have a "compelling humanitarian" justification.

From the outbreak of the Korean War in 1950, the United States had imposed fairly comprehensive economic, diplomatic, and political restrictions on North Korea. In 1999, however, President Clinton announced the United States would lift many restrictions on U.S. exports to and imports from North Korea in areas other than those controlled for national security concerns; the Departments of Commerce, Treasury, and Transportation issued new regulations a year later that implemented the new policy. On June 26, 2008, President George W. Bush delisted North Korea as a state sponsor of international terrorism, and removed restrictions based on authorities in the Trading With the Enemy Act and the terrorism designation, replacing them with more circumscribed economic restrictions related to proliferation concerns. The U.S. sanctions are a result of requirements incorporated into U.S. law by Congress, decisions made in the executive branch to exercise discretionary authorities, and obligations placed on member states of the United Nations by the U.N. Security Council. Though the President, in accordance with the Constitution, leads the way in conducting foreign policy, Congress holds substantial power to shape foreign policy by authorizing and funding programs, advising on appointments, and specifically defining the terms of engagement in accordance with U.S. political and strategic interests. This chapter presents the legislative basis for U.S. sanctions policy toward North Korea. These sanctions are a critical tenet of the larger bilateral relationship, and this chapter highlights Congress's role and responsibility in determining the nature of U.S.-North Korea relations.

Chapter 4 - Over the past 30 years, the relationship between the United States and Poland has been close and cooperative. The United States strongly supported Poland's accession to the North Atlantic Treaty Organization (NATO) in 1999 and backed its entry into the European Union (EU) in 2004. Poland has made significant contributions to U.S.- and NATO-led military operations in Iraq and Afghanistan, and Poland and the United States continue to work together closely on a range of foreign policy and international security issues. *Domestic Political and Economic Issues.* The 2015 Polish parliamentary election resulted in a victory for the conservative-nationalist Law and Justice party (PiS), which won an absolute majority of seats in the lower house of parliament (*Sejm*). Mateusz Morawiecki (PiS) is Poland's prime minister and head of government. The center-right Civic Platform (PO) party led the government of Poland from 2007 to 2015. Since winning the election, Law and Justice has made changes to the country's judicial system and enacted other reforms that have generated concerns about backsliding on democracy and triggered an EU rule-of-law investigation. Poland's next parliamentary election is due to occur in October or November 2019. European Parliament and regional election results indicate that support for Law and Justice remains strong, and the party is favored to win the 2019 election. Law and Justice candidate Andrzej Duda won Poland's 2015 presidential election. The president is Poland's head of state and exercises a number of limited but important functions. The next presidential election is due to occur in May 2020. Poland was one of the few EU economies to come through the 2008-2009 global economic crisis without major damage. As an EU member Poland is obligated to adopt the euro as its currency, but it has not set a target date for adoption and continues to use the złoty as its national currency. *Defense Modernization.* Poland has been implementing an armed forces modernization plan since 2013, and it intends to spend approximately $49 billion on military equipment acquisitions and upgrades over the period 2017-2026. Completed and prospective purchases from U.S. suppliers, including advanced Patriot missiles and F-35 Joint Strike Fighters, have a large role in this initiative. Poland is one of seven NATO members to meet the alliance's benchmark of spending at least 2% of gross domestic product (GDP) on defense, and it plans to reach 2.5% of GDP by 2030. *Defense Cooperation.* Under the United States' European Deterrence Initiative (EDI) and the U.S. military's Operation Atlantic Resolve, as well as NATO's Enhanced Forward Presence mission, U.S. forces have expanded their presence in Poland since 2014 and increased joint training and exercises with their Polish counterparts. While U.S. forces participate in these missions on a rotational basis, the Polish government has proposed the establishment of a permanent U.S. base on Polish territory. *Visa Waiver Program.* Although relations between Poland and the United States are largely positive, Poland's exclusion from the U.S. Visa Waiver Program (VWP) has been a point of contention for many years. Some Members of Congress have advocated extending the VWP to include Poland. *Relations with Russia.* Relations between Poland and Russia have long been tense, and Polish leaders have tended to view Russian intentions with wariness and suspicion. Poland remains a leading advocate for forceful EU sanctions against Russia over its 2014 annexation of Ukraine's Crimea region and fostering of separatist conflict in eastern Ukraine. *Energy Security.* Poland has promoted European energy integration, including projects to expand pipeline and electric grid interconnectivity in order to decrease reliance on Russia. Poland is a leading critic of Nord Stream 2, a Russian-owned pipeline project that would allow Germany to increase the amount of natural gas it imports directly from Russia via the Baltic Sea. *Outlook and Issues for Congress.* Given its role as a close U.S. ally and partner, Poland and its relations with the United States are of continuing congressional interest. The main areas of interest include defense cooperation, energy security, and concerns about rule-of-law and governance issues.

Chapter 5 - Venezuela remains in a deep political and economic crisis under the authoritarian rule of Nicolás Maduro of the United Socialist Party of Venezuela. Maduro, narrowly elected in 2013 after the death of Hugo Chávez (president, 1999-2013), began a second term on January 10, 2019, that most Venezuelans and much of the international community consider illegitimate. Since January, Juan Guaidó, president of Venezuela's democratically elected, opposition-controlled National Assembly, has sought to form an interim government to serve until internationally observed elections can be held. Although the United States and 53 other countries recognize Guaidó as interim president, the military high command, supported by Russia and Cuba, has remained loyal to Maduro. Venezuela is in a political stalemate as conditions in the country deteriorate. Venezuela's economy has collapsed. It is plagued by hyperinflation, severe shortages of food and medicine, and electricity blackouts that have worsened an already dire humanitarian crisis. In April 2019, United Nations officials estimated that some 90% of Venezuelans are living in poverty and 7 million need humanitarian assistance. Maduro has blamed U.S. sanctions for these problems, but most observers cite economic mismanagement and corruption under Chávez and Maduro for the current crisis. U.N. agencies estimate that 3.7 million Venezuelans had fled the country as of March 2019, primarily to other Latin American and Caribbean countries. *U.S. Policy.* The United States historically had close relations with Venezuela, a major U.S. oil supplier, but relations deteriorated under the Chávez and Maduro governments. The Trump Administration has employed targeted sanctions against Venezuelan officials responsible for human rights violations, undermining democracy, and corruption, as well as on individuals and entities engaged in drug trafficking. Since 2017, the Administration has imposed a series of broader economic sanctions on the Maduro government and state oil company Petróleos de Venezuela, S. A. (PdVSA). On January 23, 2019, the Trump Administration recognized the Guaidó government. It has imposed additional sanctions on Maduro officials, blocked Maduro and his government from benefitting from revenue from PdVSA, and imposed secondary sanctions on entities doing business with Maduro. The Administration has provided some $213 million in humanitarian assistance to countries sheltering Venezuelans and pre-positioned emergency supplies for Venezuelans at the country's borders. *Congressional Action.* The 116th Congress likely will fund foreign assistance to Venezuela and neighboring countries sheltering Venezuelans. Congress may consider additional steps to influence the Venezuelan government's behavior in promoting a return to democracy and to relieve the humanitarian crisis. In FY2019, Congress provided $17.5 million for democracy and rule of law programs in Venezuela (P.L. 116-6). The Administration's proposed FY2020 budget asks for $9 million in democracy aid and authority to transfer up to $500 million to support a transition or respond to a crisis in Venezuela. On May 20, 2019, the House Appropriations Committee reported its version of the FY2020 Department of State and Foreign Operations Appropriations Act (H.R. 2389), which would provide $17.5 million in democracy and human rights aid to Venezuela. On May 22, 2019, the Senate Foreign Relations Committee ordered S. 1025 reported with an amendment to include three House-passed measures: H.R. 854 would authorize expanded humanitarian aid, H.R. 920 would prohibit arms transfers to Maduro, and H.R. 1477 would require an assessment of Russian-Venezuelan security cooperation. S. 1025 would increase humanitarian assistance to Venezuela and countries sheltering Venezuelans and provide support for Venezuela's democratic transition. On May 22, 2019, the House Judiciary Committee ordered reported H.R. 549 to make certain Venezuelans in the United States eligible for Temporary Protected Status. On April 9, 2019, the House Committee on Foreign Affairs ordered reported H.R. 1004 to prohibit the use of U.S. military forces in combat operations in Venezuela.

Chapter 6 - More than 40 years after the end of the Vietnam War, unexploded ordnance (UXO) from numerous conflicts, but primarily dropped by U.S. forces over Cambodia, Laos, and Vietnam during the Vietnam War, continues to cause casualties in those countries. Over the past 25 years, the United States has provided a total of over $400 million in assistance for UXO clearance and related activities in those three countries through the Department of Defense (DOD), Department of State (DOS), and United States Agency for International Development (USAID), as well as funding for treatment of victims through USAID and the Leahy War Victims fund. Although casualty numbers have dropped in recent years, no systematic assessment of affected areas has been done, and many observers believe it may still take decades to clear the affected areas. War legacy issues such as UXO clearance and victim assistance may raise important considerations for Congress as it addresses the impact of U.S. participation in conflicts around the world and how the United States should deal with the aftermath of such conflicts. The continued presence of UXO in Southeast Asia raises numerous issues, including appropriate levels of U.S. assistance for clearance activities and victim relief; coordination in efforts among DOD, DOS, and USAID; the implications of U.S. action on relations with affected countries; whether U.S. assistance in Southeast Asia carries lessons for similar activity in other parts of the world, including Iraq and Afghanistan; and, more generally, efforts to lessen the prevalence of UXO in future conflicts. Many observers argue that U.S. efforts to address UXO issues in the region, along with joint efforts regarding other war legacy issues such as POW/MIA identification and Agent Orange/dioxin remediation, have been important steps in building relations with the affected countries in the post-war period. These efforts that have proceeded furthest in Vietnam, where the bilateral relationship has expanded across a wide range of economic and security initiatives. In Cambodia and Laos, where bilateral relations are less developed, UXO clearance is one of the few issues on which working-level officials from the United States and the affected countries have cooperated for years. Although some Cambodians and Laotians view U.S. demining assistance as a moral obligation and the U.S. government has viewed its support for UXO clearance as an important, positive aspect of its ties with the two countries, the issue of UXO has not been a major factor driving the relationships. The Consolidated Appropriations Act, 2019 (P.L. 116-6) provides $196.5 million for "conventional weapons destruction" around the world, including $159.0 million for "humanitarian demining," with $3.85 million appropriated for Cambodia, $30.0 million for Laos, and $15.0 million for Vietnam. The Legacies of War Recognition and Unexploded Ordnance Removal Act (H.R. 2097) would authorize $50 million per year for fiscal years 2020 to 2024 for humanitarian assistance in Cambodia, Laos, and Vietnam to develop national UXO surveys, conduct UXO clearance, and finance capacity building, risk education, and support for UXO victims.

In: Key Congressional Reports for June 2019 ISBN: 978-1-53616-666-8
Editor: Tara T. Robertson

Chapter 1

CHINA'S ECONOMIC RISE: HISTORY, TRENDS, CHALLENGES, AND IMPLICATIONS FOR THE UNITED STATES*

Wayne M. Morrison

ABSTRACT

Prior to the initiation of economic reforms and trade liberalization nearly 40 years ago, China maintained policies that kept the economy very poor, stagnant, centrally controlled, vastly inefficient, and relatively isolated from the global economy. Since opening up to foreign trade and investment and implementing free-market reforms in 1979, China has been among the world's fastest-growing economies, with real annual gross domestic product (GDP) growth averaging 9.5% through 2018, a pace described by the World Bank as "the fastest sustained expansion by a major economy in history." Such growth has enabled China, on average, to double its GDP every eight years and helped raise an estimated 800 million people out of poverty. China has become the world's largest economy (on a purchasing power parity basis),

* This is an edited, reformatted and augmented version of Congressional Research Service, Publication No. RL33534, dated June 25, 2019.

manufacturer, merchandise trader, and holder of foreign exchange reserves. This in turn has made China a major commercial partner of the United States. China is the largest U.S. merchandise trading partner, biggest source of imports, and third-largest U.S. export market. China is also the largest foreign holder of U.S. Treasury securities, which help fund the federal debt and keep U.S. interest rates low.

As China's economy has matured, its real GDP growth has slowed significantly, from 14.2% in 2007 to 6.6% in 2018, and that growth is projected by the International Monetary Fund (IMF) to fall to 5.5% by 2024. The Chinese government has embraced slower economic growth, referring to it as the "new normal" and acknowledging the need for China to embrace a new growth model that relies less on fixed investment and exporting, and more on private consumption, services, and innovation to drive economic growth. Such reforms are needed in order for China to avoid hitting the "middle-income trap," when countries achieve a certain economic level but begin to experience sharply diminishing economic growth rates because they are unable to adopt new sources of economic growth, such as innovation.

The Chinese government has made innovation a top priority in its economic planning through a number of high-profile initiatives, such as "Made in China 2025," a plan announced in 2015 to upgrade and modernize China's manufacturing in 10 key sectors through extensive government assistance in order to make China a major global player in these sectors. However, such measures have increasingly raised concerns that China intends to use industrial policies to decrease the country's reliance on foreign technology (including by locking out foreign firms in China) and eventually dominate global markets.

In 2017, the Trump Administration launched a Section 301 investigation of China's innovation and intellectual property policies deemed harmful to U.S. economic interests. It subsequently raised tariffs by 25% on $250 billion worth of imports from China, while China increased tariffs (ranging from 5% to 25%) on $110 billion worth of imports from the United States. Such measures have sharply decreased bilateral trade in 2019. On May 10, 2019, President Trump announced he was considering raising tariffs on nearly all remaining products from China. A protracted and escalating trade conflict between the United States and China could have negative consequences for the Chinese economy.

China's growing global economic influence and the economic and trade policies it maintains have significant implications for the United States and hence are of major interest to Congress. While China is a large and growing market for U.S. firms, its incomplete transition to a free-market economy has resulted in economic policies deemed harmful to U.S. economic interests, such as industrial policies and theft of U.S. intellectual property. This chapter provides background on China's economic rise; describes its current economic structure; identifies the challenges China faces to maintain economic growth; and discusses the challenges, opportunities, and implications of China's economic rise for the United States.

INTRODUCTION

China's rise from a poor developing country to a major economic power in about four decades has been spectacular. From 1979 (when economic reforms began) to 2017, China's real gross domestic product (GDP) grew at an average annual rate of nearly 10%.[1] According to the World Bank, China has "experienced the fastest sustained expansion by a major economy in history—and has lifted more than 800 million people out of poverty."[2] China has emerged as a major global economic power. For example, it ranks first in terms of economic size on a purchasing power parity (PPP) basis, value-added manufacturing, merchandise trade, and holder of foreign exchange reserves.

China's rapid economic growth has led to a substantial increase in bilateral commercial ties with the United States. According to U.S. trade data, total trade between the two countries grew from $5 billion in 1980 to $660 billion in 2018. China is currently the United States' largest merchandise trading partner, its third-largest export market, and its largest source of imports. Many U.S. companies have extensive operations in China in order to sell their products in the booming Chinese market and to take advantage of lower-cost labor for export-oriented manufacturing.[3] These operations have helped some U.S. firms to remain internationally competitive and have supplied U.S. consumers with a variety of low-cost

---

[1] China's economic reform process began in December 1978 when the Third Plenum of the Eleventh Central Committee of the Communist Party adopted Deng Xiaoping's economic proposals. Implementation of the reforms began in 1979.

[2] World Bank, *China Overview*, March 28, 2017, available at http://www.worldbank.org/en/country/china/overview.

[3] Some companies use China as part of their global supply chain for manufactured parts, which are then exported and assembled elsewhere. Other firms have shifted the production of finished products from other countries (mainly in Asia) to China; they import parts and materials into China for final assembly.

goods. China's large-scale purchases of U.S. Treasury securities (which totaled $1.1 trillion as of April 2019 have enabled the federal government to fund its budget deficits, which help keep U.S. interest rates relatively low.[4]

However, the emergence of China as a major economic power has raised concern among many U.S. policymakers. Some claim that China uses unfair trade practices (such as an undervalued currency and subsidies given to domestic producers) to flood U.S. markets with low-cost goods, and that such practices threaten American jobs, wages, and living standards. Others contend that China's growing use of industrial policies to promote and protect certain domestic Chinese industries or firms favored by the government, and its failure to take effective action against widespread infringement and theft of U.S. intellectual property rights (IPR) in China, threaten to undermine the competitiveness of U.S. IP-intensive industries. In addition, while China has become a large and growing market for U.S. exports, critics contend that numerous trade and investment barriers limit opportunities for U.S. firms to sell in China, or force them to set up production facilities in China as the price of doing business there.

The Chinese government views a growing economy as vital to maintaining social stability. However, China faces a number of major economic challenges that could dampen future growth, including distortive economic policies that have resulted in overreliance on fixed investment and exports for economic growth (rather than on consumer demand), government support for state-owned firms, a weak banking system, widening income gaps, growing pollution, and the relative lack of the rule of law in China. The Chinese government has acknowledged these problems and has pledged to address them by implementing policies to increase the role of the market in the economy, boost innovation, make consumer spending the driving force of the economy, expand social safety net coverage, encourage the development of less-polluting industries (such as services), and crack down on official government corruption. The ability

---

[4] See CRS Report RL33536, *China-U.S. Trade Issues*, by Wayne M. Morrison.

of the Chinese government to implement such reforms will likely determine whether China can continue to maintain relatively rapid economic growth rates, or will instead begin to experience significantly lower growth rates.

China's growing economic power has led it to become increasingly involved in global economic policies and projects, especially infrastructure development. China's Belt and Road initiative (BRI) represents a grand strategy by China to finance infrastructure throughout Asia, Europe, Africa, and beyond. If successful, China's economic initiatives could significantly expand export and investment markets for China and increase its "soft power" globally.

This chapter provides background on China's economic rise; describes its current economic structure; identifies the challenges China faces to maintain economic growth; and discusses the challenges, opportunities, and implications of China's economic rise for the United States.

THE HISTORY OF CHINA'S ECONOMIC DEVELOPMENT

China's Economy Prior to Reforms

Prior to 1979, China, under the leadership of Chairman Mao Zedong, maintained a centrally planned, or command, economy. A large share of the country's economic output was directed and controlled by the state, which set production goals, controlled prices, and allocated resources throughout most of the economy. During the 1950s, all of China's individual household farms were collectivized into large communes. To support rapid industrialization, the central government undertook large-scale investments in physical and human capital during the 1960s and 1970s. As a result, by 1978 nearly three-fourths of industrial production was produced by centrally controlled, state-owned enterprises (SOEs), according to centrally planned output targets. Private enterprises and foreign-invested firms were generally barred. A central goal of the Chinese government was to make China's economy relatively self-sufficient.

Foreign trade was generally limited to obtaining those goods that could not be made or obtained in China. Such policies created distortions in the economy. Since most aspects of the economy were managed and run by the central government, there were no market mechanisms to efficiently allocate resources, and thus there were few incentives for firms, workers, and farmers to become more productive or be concerned with the quality of what they produced (since they were mainly focused on production goals set by the government).

According to Chinese government statistics, China's real GDP grew at an average annual rate of 6.7% from 1953 to 1978, although the accuracy of these data has been questioned by many analysts, some of whom contend that during this period, Chinese government officials (especially at the subnational levels) often exaggerated production levels for a variety of political reasons. Economist Angus Maddison puts China's actual average annual real GDP growth during this period at about 4.4%.[5] In addition, China's economy suffered significant economic downturns during the leadership of Chairman Mao Zedong, including during the Great Leap Forward from 1958 to 1962 (which led to a massive famine and reportedly the deaths of up to 45 million people)[6] and the Cultural Revolution from 1966 to 1976 (which caused widespread political chaos and greatly disrupted the economy). From 1950 to 1978, China's per capita GDP on a purchasing power parity (PPP) basis,[7] a common measurement of a country's living standards, doubled. However, from 1958 to 1962, Chinese living standards fell by 20.3%, and from 1966 to 1968, they dropped by 9.6% (see Figure 1). In addition, the growth in Chinese living standards paled in comparison to those in the West, such as Japan, as indicated in Figure 2.

---

[5] The Organization for Economic Cooperation and Development, *Chinese Economic Performance in the Long Run, 960-2030*, by Angus Maddison, 2007.

[6] New York Times, Editorial, *Mao's Great Leap to Famine*, December 15, 2010.

[7] Purchasing power parities are a method used to measure and compare the economic data of other countries expressed in U.S. dollars. That method adjusts the data to reflect differences in prices across countries. This method is discussed in more detail later in the report.

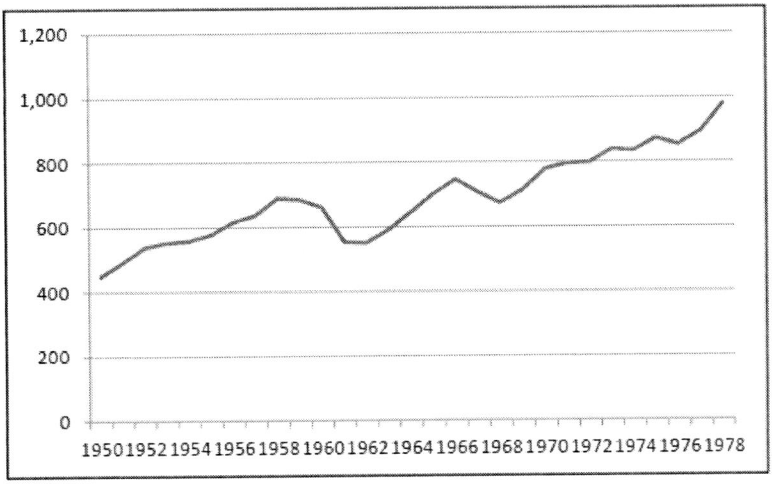

Source: Angus Maddison, Historical, Statistics of the World Economy: 1-2008 AD.

Figure 1. Chinese per capita GDP: 1950-1978 ($ billions, PPP basis).

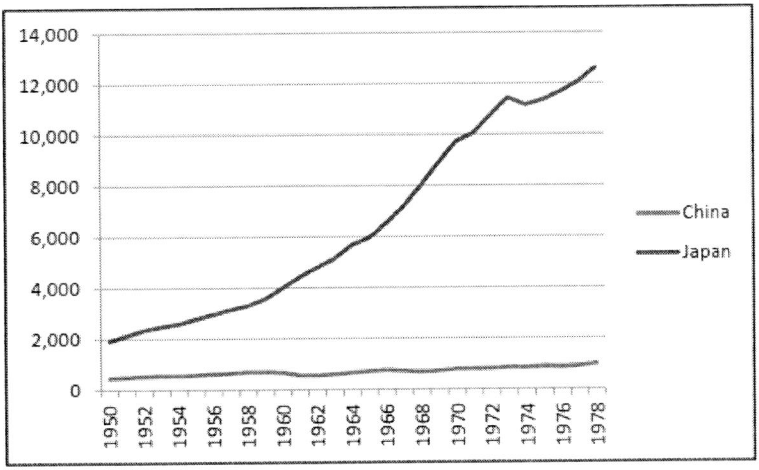

Source: Angus Maddison, Historical, Statistics of the World Economy: 1-2008 AD.

Figure 2. Comparison of Chinese and Japanese per capita GDP: 1950-1978 ($ billions, PPP basis).

In 1978, (shortly after the death of Chairman Mao in 1976), the Chinese government decided to break with its Soviet-style economic policies by gradually reforming the economy according to free market principles and opening up trade and investment with the West, in the hope that this would significantly increase economic growth and raise living standards. As Chinese leader Deng Xiaoping, the architect of China's economic reforms, put it: "Black cat, white cat, what does it matter what color the cat is as long as it catches mice?"[8]

The Introduction of Economic Reforms

Beginning in 1979, China launched several economic reforms. The central government initiated price and ownership incentives for farmers, which enabled them to sell a portion of their crops on the free market. In addition, the government established four special economic zones along the coast for the purpose of attracting foreign investment, boosting exports, and importing high technology products into China. Additional reforms, which followed in stages, sought to decentralize economic policymaking in several sectors, especially trade. Economic control of various enterprises was given to provincial and local governments, which were generally allowed to operate and compete on free market principles, rather than under the direction and guidance of state planning. In addition, citizens were encouraged to start their own businesses. Additional coastal regions and cities were designated as open cities and development zones, which allowed them to experiment with free-market reforms and to offer tax and trade incentives to attract foreign investment. In addition, state price controls on a wide range of products were gradually eliminated. Trade liberalization was also a major key to China's economic success. Removing trade barriers encouraged greater competition and attracted FDI inflows. China's gradual implementation of economic reforms sought to identify which policies produced favorable economic outcomes (and which did not) so that they could be implemented in other parts of the country, a

---

[8] This reference appears to have meant that it did not matter whether an economic policy was considered to be "capitalist" or "socialist," what really mattered was whether that policy would boost the economy and living standards.

process Deng Xiaoping reportedly referred to as "crossing the river by touching the stones."[9]

China's Economic Growth and Reforms: 1979-the Present

Since the introduction of economic reforms, China's economy has grown substantially faster than during the pre-reform period, and, for the most part, has avoided major economic disruptions.[10] From 1979 to 2018, China's annual real GDP averaged 9.5% (see Figure 3). This has meant that on average China has been able to double the size of its economy in real terms every eight years. The global economic slowdown, which began in 2008, had a significant impact on the Chinese economy. China's media reported in early 2009 that 20 million migrant workers had returned home after losing their jobs because of the financial crisis and that real GDP growth in the fourth quarter of 2008 had fallen to 6.8% year-on-year. The Chinese government responded by implementing a $586 billion economic stimulus package, aimed largely at funding infrastructure and loosening monetary policies to increase bank lending.[11] Such policies enabled China to counter the effects of the sharp global fall in demand for Chinese products. From 2008 to 2010, China's real GDP growth averaged 9.7%. However, the rate of GDP growth declined slowed for the next six consecutive years, falling from 10.6% in 2010 to 6.7% in 2016. Real GDP ticked up to 6.8% in 2017, but slowed to 6.6% in 2018, (although it rose to 6.8% in 2017).

---

[9] Many analysts contend that Deng's push to implement economic reforms was largely motivated by a belief that they would boost economic growth and thus strengthen the power of the Chinese Communist Party.

[10] China's economic growth slowed significantly followed the aftermath of the Tiananmen massacre that occurred in June 1989. Several countries, including the United States, imposed trade sanctions against China, and Chinese economic reforms were essentially put on hold. China's real GDP growth rate fell from 11.3% in 1988 to 4.2% in 1989, and declined to 3.9% in 1990. In 1991, economic reforms were restarted and foreign sanctions against China were reduced or removed, and real GDP grew by 9.2%.

[11] Xinhuanet, "20 million jobless migrant workers return home," February 2, 2009.

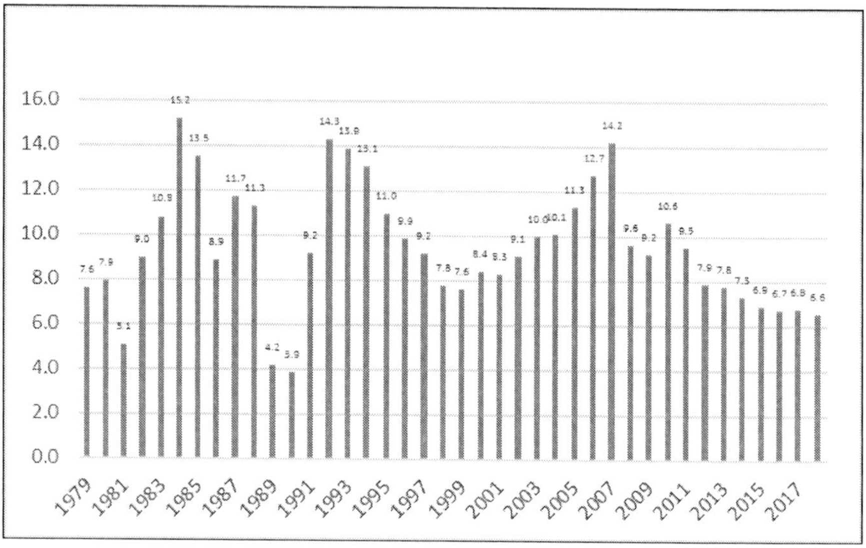

Source: IMF, and Chinese National Bureau of Statistics.

Figure 3. Chinese Annual Real GDP Growth: 1979-2018 (percentage change).

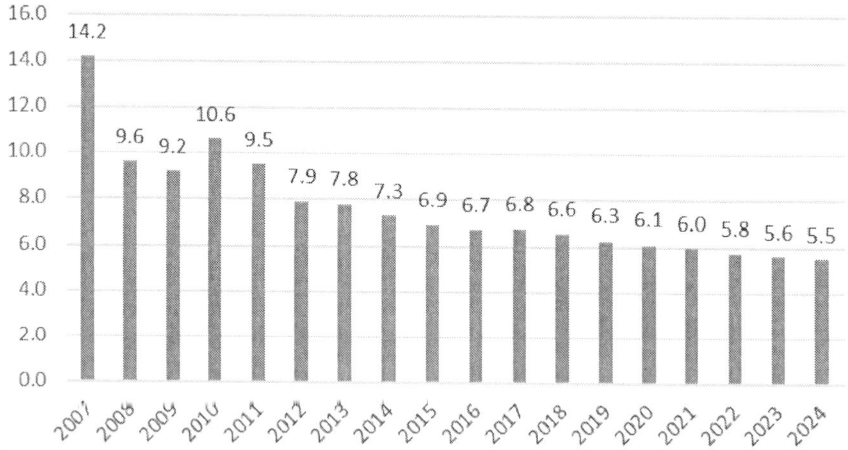

Source: IMF, World Economic Outlook Database, April 2019.

Figure 4. China's Real Annual GDP Growth: 2007-2018 and Projections through 2024 (percentage).

The IMF's April 2019 World Economic Outlook projects that China's real GDP growth will slow each year over the next six years, falling to 5.5% in 2024 (Figure 4).[12] Many economists warn that China's economic growth could slow further if the United States and China continue to impose punitive economic measures against each other, such the tariff hikes that have resulted from U.S. action under Section 301 and Chinese retaliation. The Organization for Economic and Cooperation and Development (OECD) projects that increased tariffs on all trade between the United States and China could reduce China's real GDP in 2021-2022 by 1.1% relative to the OECD's baseline economic projections.[13]

Causes of China's Economic Growth

Economists generally attribute much of China's rapid economic growth to two main factors: large-scale capital investment (financed by large domestic savings and foreign investment) and rapid productivity growth. These two factors appear to have gone together hand in hand.

Economic reforms led to higher efficiency in the economy, which boosted output and increased resources for additional investment in the economy.

China has historically maintained a high rate of savings. When reforms were initiated in 1979, domestic savings as a percentage of GDP stood at 32%. However, most Chinese savings during this period were generated by the profits of SOEs, which were used by the central government for domestic investment. Economic reforms, which included the decentralization of economic production, led to substantial growth in Chinese household savings as well as corporate savings. As a result, China's gross savings as a percentage of GDP is the highest among major economies. The large level of domestic savings has enabled China to

---

[12] IMF, *World Economic Outlook Database,* April 2019.

[13] OECD, Economic Outlook, May 2019, available at https://www.oecd-ilibrary.org/docserver/b2e897b0-en.pdf?expires=1561458758&id=id&accname=oid011901&checksum=40A52BB1E685ADAB80433EDD227A4D65.

support a high level of investment. In fact, China's gross domestic savings levels far exceed its domestic investment levels, which have made China a large net global lender.

Several economists have concluded that productivity gains (i.e., increases in efficiency) have been another major factor in China's rapid economic growth. The improvements to productivity were caused largely by a reallocation of resources to more productive uses, especially in sectors that were formerly heavily controlled by the central government, such as agriculture, trade, and services. For example, agricultural reforms boosted production, freeing workers to pursue employment in the more productive manufacturing sector. China's decentralization of the economy led to the rise of non-state enterprises (such as private firms), which tended to pursue more productive activities than the centrally controlled SOEs and were more market-oriented and more efficient. Additionally, a greater share of the economy (mainly the export sector) was exposed to competitive forces. Local and provincial governments were allowed to establish and operate various enterprises without interference from the government. In addition, FDI in China brought with it new technology and processes that boosted efficiency.

However, as China's technological development begins to converge with major developed countries (i.e., through its adoption of foreign technology), its level of productivity gains, and thus, real GDP growth, could slow significantly from its historic levels unless China becomes a major center for new technology and innovation and/or implements new comprehensive economic reforms. Several developing economies (notably several in Asia and Latin America) experienced rapid economic development and growth during the 1960s and 1970s by implementing some of the same policies that China has utilized to date to develop its economy, such as measures to boost exports and to promote and protect certain industries. However, at some point in their development, some of these countries began to experience economic stagnation (or much slower growth compared to previous levels) over a sustained time, a phenomenon described by economists as the "middle-income trap."[14] This means that several developing (low-income) economies were able to transition to a middle-income economy, but because they were unable to sustain high levels of productivity gains (in part, because they could not address structural inefficiencies in the economy), they were unable to transition to a high-income economy.[15] China may be at a similar crossroads now. The World Bank classifies development levels of economies using a per capita gross national income (GNI) methodology.[16] According to the World Bank, China went from a low-income economy to a low-middle-income economy in 1997, and in 2010, it became an upper-middle-income country. China's 2017 per capita GNI (at $8,690) was 38.7% below the level China would need to obtain to become a high-income economy. The Chinese government projects that China can cross the high-income threshold by 2025. It hopes to achieve this largely by making innovation a major source of future economic growth. Skeptics contend that innovation growth in China will be hard to achieve, especially if it is mainly state-driven and imposes new restrictions on foreign firms,

The Economist Intelligence Unit (EIU) projects that China's real GDP growth will slow considerably over the next several decades, eventually converging on U.S. growth rates by the year 2037 (U.S. and Chinese real GDP growth rates are both projected at 1.9%). For some years thereafter, EIU projects U.S. GDP growth to be greater than China's (Figure 6).[17]

The Chinese government has indicated its desire to move away from its current economic model of fast growth at any cost to more "smart" economic growth, which seeks to reduce reliance on energy-intensive and high-polluting industries and rely more on high technology, green energy, and services. China also has indicated it wants to obtain more balanced economic growth. (These issues are discussed in more detail later in the report).

---

[14] Japan was able to become a high-income economy, but since the mid-1980s, its economic growth has been relatively stagnant.

[15] These designations are based on World Bank per capita GDP measurements.

[16] The classifications are determined by per capita income ranges (the thresholds of which are adjusted annually). These include low-income economies, lower-middle-income economies, upper-middle-income countries, and high-income countries.

[17] Long-term economic projections should be interpreted with caution.

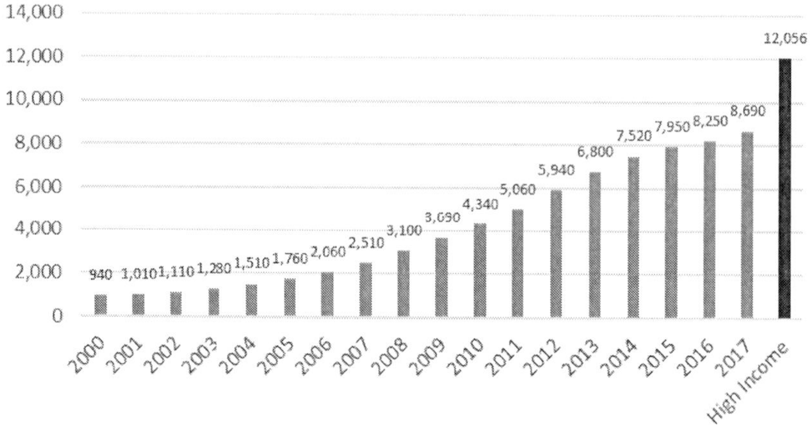

Source: World Bank.

Notes: Bar in red indicates the level China would need to reach to become a high-income economy.

Figure 5. World Bank Measurements of China's Per Capita GNI: 2000-2017 ($ U.S.).

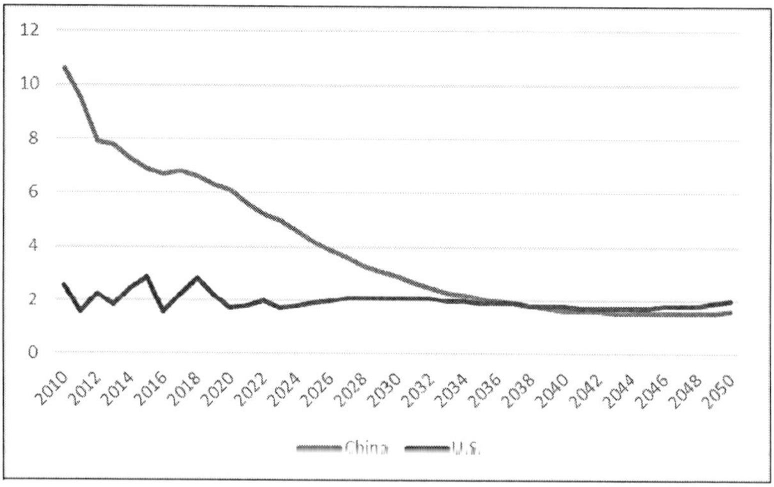

Source: EIU Database (accessed on June 24, 2019).

Figure 6. U.S. and Chinese Annual Real GDP Growth Rates in 2010-2018 and Projections through 2050 (percentage).

MEASURING THE SIZE OF CHINA'S ECONOMY

The rapid growth of the Chinese economy has led many analysts to speculate if and when China will overtake the United States as the "world's largest economic power." The "actual" size of China's economy has been a subject of extensive debate among economists. Measured in U.S. dollars using nominal exchange rates, China's GDP in 2018 in nominal U.S. dollars was $13.4 trillion, which was 65.3% of the size of the U.S. economy, according to estimates made by the IMF. China's 2018 per capita GDP in nominal dollars was $9,608, which was 15.3% of the U.S. per capita level.

Many economists contend that using nominal exchange rates to convert Chinese data (or those of other countries) into U.S. dollars fails to reflect the true size of China's economy and living standards relative to the United States. Nominal exchange rates simply reflect the prices of foreign currencies vis-à-vis the U.S. dollar, and such measurements exclude differences in the prices for goods and services across countries. To illustrate, one U.S. dollar exchanged for local currency in China would buy more goods and services there than it would in the United States. This is because prices for goods and services in China are generally lower than they are in the United States. Conversely, prices for goods and services in Japan are generally higher than they are in the United States (and China). Thus, one dollar exchanged for local Japanese currency would buy fewer goods and services there than it would in the United States. Economists attempt to develop estimates of exchange rates based on their actual purchasing power relative to the dollar in order to make more accurate comparisons of economic data across countries, usually referred to as purchasing power parity (PPP).

The PPP exchange rate increases the (estimated) measurement of China's economy and its per capita GDP. According to the IMF (which uses price surveys conducted by the World Bank), prices for goods and services in China are about half the level they are in the United States. Adjusting for this price differential raises the value of China's 2018 GDP from $13.4 trillion (nominal dollars) to $25.3 trillion (on a PPP basis) (see

Table 1).[18] IMF data indicate that China overtook the United States as the world's largest economy in 2014 on a PPP basis.[19]

China's share of global GDP on a PPP basis rose from 2.3% in 1980 to an estimated 18.3% in 2017, while the U.S. share of global GDP on a PPP basis fell from 24.3% to an estimated 15.3%.[20] This would not be the first time in history that China was the world's largest economy (see text box). China's economic ascendency as the world largest economy has been impressive, especially considering that in 1980, China's GDP on a PPP basis was only one-tenth that of the United States (see Figure 7). The IMF predicts that by 2024, China's economy will be 56% larger than the U.S. economy on a PPP basis.

Table 1. Comparisons of Chinese, Japanese, and U.S. GDP and per capita GDP in nominal U.S. dollars and a purchasing power parity basis: 2018

	China	United States
Nominal GDP ($ billions)	13,407	20,494
GDP in PPP ($ billions)	25,270	20,494
Nominal Per Capita GDP ($)	9,608	62,606
Per Capita GDP in PPP ($)	18,110	62,606

Source: IMF, World Economic Forum.

The Decline and Rise of China's Economy

According to a study by economist Angus Maddison, China was the world's largest economy in 1820, accounting for an estimated 32.9% of global GDP. However, foreign and civil wars, internal strife, weak and ineffective governments, natural disasters (some of which were man-made), and distortive economic policies caused China's share of global GDP on a PPP basis to shrink significantly. By 1952, China's share of global GDP had fallen to 5.2%, and by 1978, it slid to 4.9%. The adoption of economic reforms by China in the late 1970s led to a surge in China's economic growth and helped restore China as a major global economic power.

Source: The Organization for Economic Cooperation and Development, *Chinese Economic Performance in the Long Run, 960-2030*, by Angus Maddison, 2007.

---

[18] PPP data reflect what the value of China's goods and services would be if they were sold in the United States.

[19] The United States remains the world's largest economy when using nominal U.S. dollars.

[20] IMF, World Economic Outlook, October 2017, projections

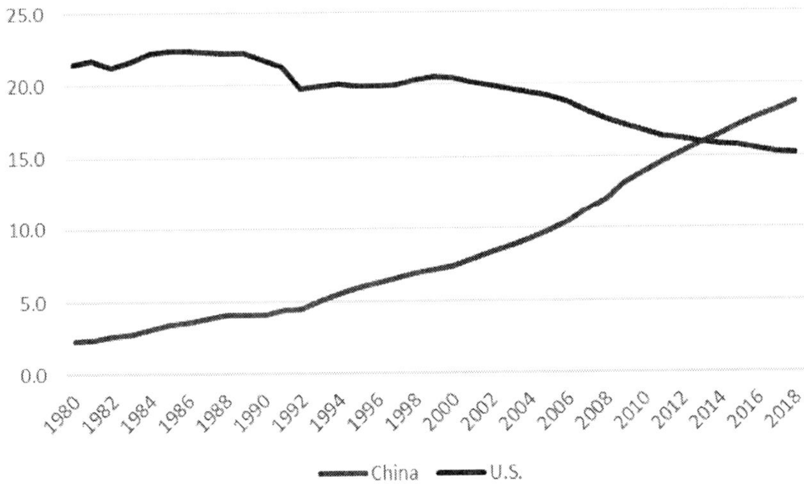

Source: IMF, World Economic Outlook, April 2019.

Figure 7. U.S. and Chinese GDP (PPP Basis) as a Share of Global Total: 1980-2018 (%).

The PPP measurement also raises China's 2018 nominal per capita GDP (from $9,608) to $18,110, which was 28.9% of the U.S. level. Even with continued rapid economic growth, it would likely take many years for Chinese living standards to approach U.S. levels.

China as the World's Largest Manufacturer

China has emerged as the world's largest manufacturer according to the World Bank. Figure 8 lists estimates of the gross value added of manufacturing in China, the United States, and Japan expressed in U.S. dollars in 2006 and 2016. Gross value added data reflect the actual value of manufacturing that occurred in the country (i.e., they subtract the value of intermediate inputs and raw materials used in production). In 2016, the value of China's manufacturing on a gross value added basis was 49.2% higher than the U.S. level. Manufacturing plays a considerably more important role in the Chinese economy than it does for the United States.

In 2016, China's gross valued added manufacturing was equal to 28.7% of its GDP, compared to 11.6% for the United States.[21]

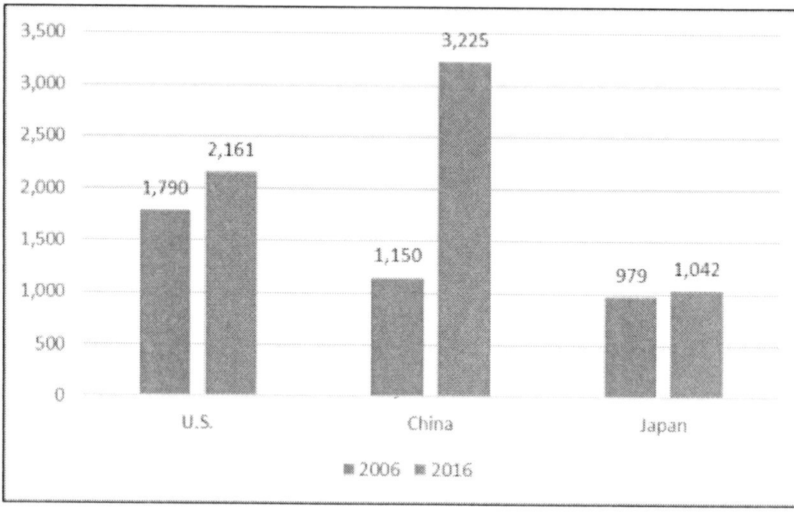

Source: The World Bank.

Figure 8. Gross Value Added Manufacturing in China, the United States, and Japan: 2006 and 2016 ($ billions).

In its 2016 Global Manufacturing Competitiveness Index, Deloitte (an international consulting firm) ranked China as the world's most competitive manufacturer (out of 40 countries), based on a survey of global manufacturing executives, while the United States ranked second (it ranked fourth in 2010). The index found that global executives predicted that the United States would overtake China by 2020 to become the world's most competitive economy, largely because of its heavy investment in talent and technology (e.g., high levels of R&D spending and activities, the presence of top-notch universities, and large amounts of venture capital being invested in advanced technologies). On the other hand, while China was expected to remain a major manufacturing power because of its large R&D spending levels, movement toward higher-valued, advanced manufacturing, government policies to promote innovation, and a large

[21] The World Bank, *Data*, at https://data.worldbank.org/indicator/NV.IND.MANF.CD.

pool of graduates in science, technology, engineering and mathematics, it was viewed as facing several challenges, including a slowing economy, a decline in value-added manufacturing and overcapacity in several industries, rising labor costs, and a rapidly aging population. As a result, China was projected to fall to the second-most competitive manufacturer by 2020.[22]

More broadly, the *World Economic Forum* (WEF) produces an annual report that assesses and ranks (based on an index) the global competitiveness of a country's entire economy, based on factors that determine the level of productivity of an economy, which in turn sets the level of prosperity that the country can achieve. The WEF's 2018 Global Competitive Index ranked China as the world's 28th-most competitive economy (out of 140 economies), while the United States ranked first.[23]

Changes in China's Wage and Labor Cost Advantages

The decline in China's working age population may have contributed rising wages in China. As indicated in Figure 9, China's average monthly wages (converted into U.S. dollars) in 1990 were $55, compared with $32 for Vietnam and $221 for Mexico.[24] However, in 2018, China's average monthly wages (at $990) were 316% higher than Vietnam's wages ($238) and 158.5% higher than Mexico's ($383). From 2007 to 2018, China's average monthly wages rose by 263%. The American Chamber of Commerce in China (AmCham China) 2019 Business Climate survey listed rising labor costs as the second-biggest challenge facing U.S. firms in China (56% of recipients cited them as their largest concern).[25]

---

[22] Deloitte, *2016 Global Manufacturing Competitiveness Index*, 2016, available at https://www2.deloitte.com/content/dam/Deloitte/us/Documents/manufacturing/us-gmci.pdf.

[23] World Economic Forum, *The Global Competitiveness Report, 2016–2017*, September 2016.

[24] The Economist Intelligence Unit, *Data Tool, accessed in June 2019.*

[25] AmCham China, *2018 Business Climate Survey Report*, January 2017, available at http://www.amchamchina.org/.

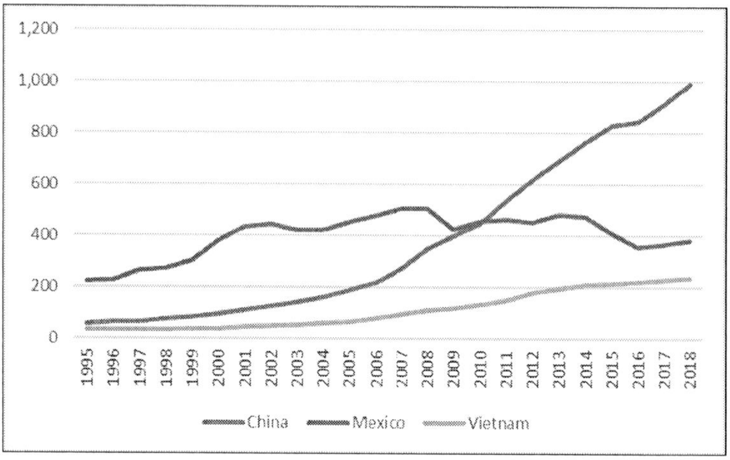

Source: Economist Intelligence Unit.

Notes: Because data are listed in U.S. dollars rather than local currency, changes to wages may also partially reflect changes to exchange rates with the U.S. dollar. However, such data may reflect average labor costs in dollars that U.S.-invested firms might face in their overseas operations.

Figure 9. Average Monthly Wages for China, Mexico and Vietam: 1990-2018. (Nominal U.S. dollars).

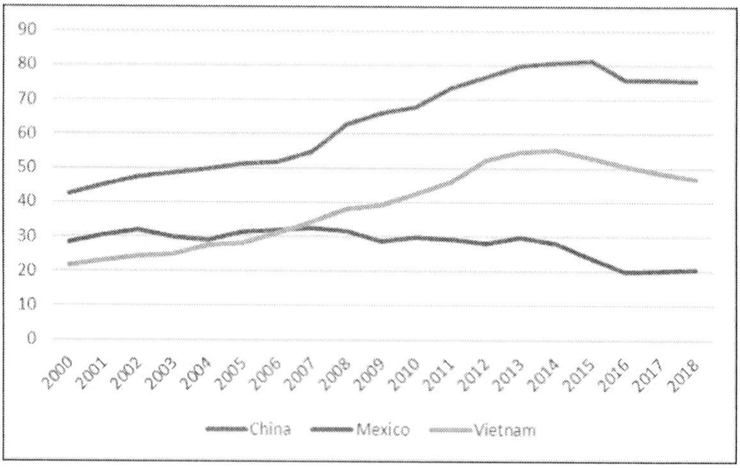

Source: Economist Intelligence Unit.

Notes: The labor cost of producing one unit of output, indexed to U.S. levels.

Figure 10. Labor Cost Index for China, Mexico, and Vietnam Relative to those in the United States: 2000-2018 (U.S. level = 100).

Figure 10 shows a comparison of labor costs per unit of production for the countries listed in the previous figure, indexed relative to U.S levels. In 2000, China's unit labor production costs were 42.3 of U.S. levels and by 2018 they rose to 75.5% of U.S. levels.[26]

FOREIGN DIRECT INVESTMENT (FDI) IN CHINA

China's trade and investment reforms and incentives led to a surge in FDI beginning in the early 1990s. Such flows have been a major source of China's productivity gains and rapid economic and trade growth. There were reportedly 445,244 foreign-invested enterprises (FIEs) registered in China in 2010, employing 55.2 million workers or 15.9% of the urban workforce.[27] As indicated in Figure 11, FIEs account for a significant share of China's industrial output. That level rose from 2.3% in 1990 to a high of 35.9% in 2003, but fell to 25.9% in 2011.[28] In addition, FIEs are responsible for a significant level of China's foreign trade. At their peak, FIEs accounted for 58.3% of Chinese exports in 2005 and 59.7% of imports, but these levels have subsequently fallen, reaching 41.7% and 43.7%, respectively, in 2018 (see Figure 12).

The United Nations Conference on Trade and Development (UNCTAD) reports that China has become a both a major recipient of global FDI as well as a major provider of FDI outflows (see Figure 13).[29] China's FDI inflows in 2018 were $139 billion, making it the world's second-largest recipient of FDI after the United States.[30]

---

[26] The Economist Intelligence Unit, Data Tool, accessed in June 2019.

[27] China 2012 Statistical Yearbook.

[28] Industrial output is defined by the Chinese government as the total volume of final industrial products produced and industrial services provided during a given period. Source: China 2012 Statistical Yearbook.

[29] U.N.FDI data differ from Chinese data, in part because Chinese data are limited to nonfinancial FDI and UN data includes financial-related FDI. UNCTAD reports Hong Kong FDI data separately.

[30] UNCTAD, *2019 World Investment Report*, available at https://unctad.org/en/PublicationsLibrary/wir2019_en.pdf.

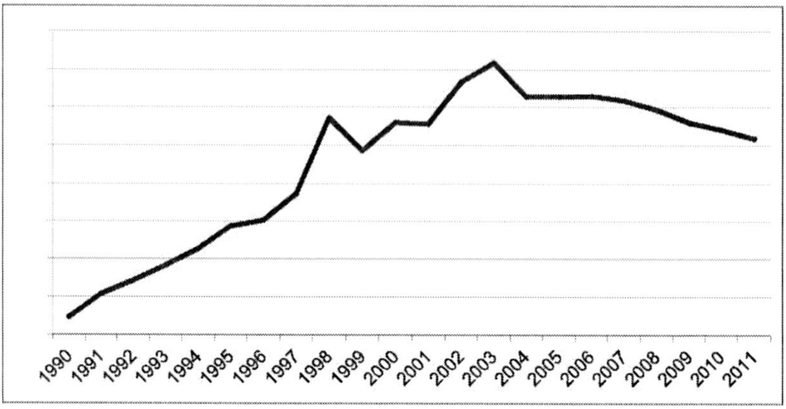

Source: Invest in China (http://www.fdi.gov.cn) and China's 2012 Statistical Yearbook.

Figure 11. Industrial Output by Foreign-Invested Firms in China as a Share of National Output Total: 1990-2011 (percentage).

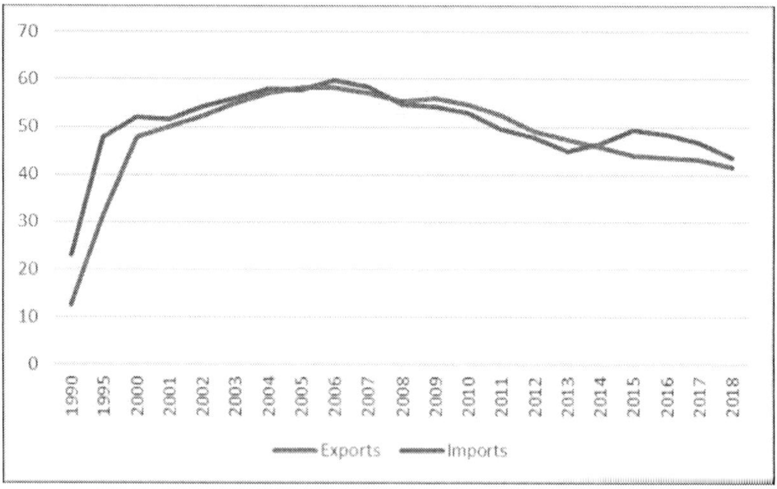

Source: Invest in China (http://www.fdi.gov.cn).

Figure 12. Share of Chinese Merchandise Exports and Imports by Foreign-Invested Enterprises in China: 1990-2018. (percentage).

China's FDI outflows have grew rapidly after 2005 and exceeded FDI inflows for the first time in 2015. China's FDI outflows reached a historic peak of $196.1 billion in 2016, but declined in 2017 and 2018, reflecting a crackdown by the Chinese government on investment deemed wasteful and

well as greater scrutiny by foreign governments of China's efforts to obtain advanced technology firms and other strategic assets. Still, China was the world's second-largest source of FDI outflows (after Japan).

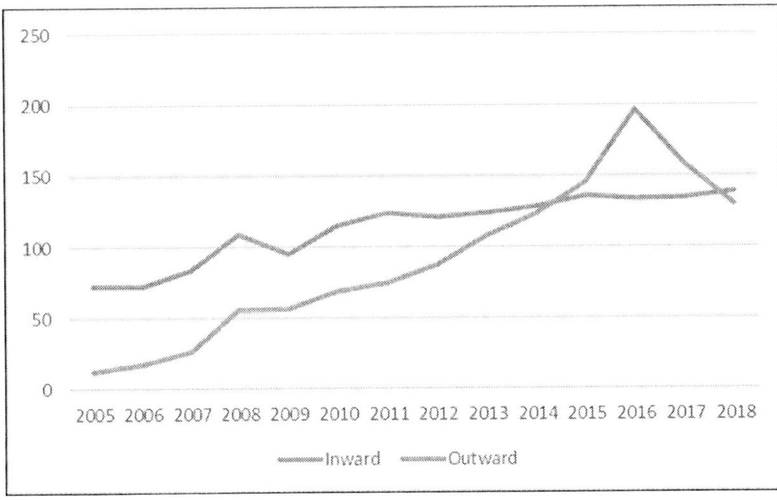

Source: UNCTAD.
Notes: UNCTAD FDI data differ from that reported by China.

Figure 13. Estimates of China's Annual FDI Inflows and Outflows: 2005-2018 ($ billions).

The sharp increase in China's global FDI outflows in recent years appears to be largely driven by a number of factors, including Chinese government policies and initiatives to encourage firms to "go global." The government wants to use FDI to gain access to IPR, technology, know-how, famous brands, etc., in order to move Chinese firms up the value-added chain in manufacturing and services, boost domestic innovation and development of Chinese brands, and help Chinese firms (especially SOEs) to become major global competitors.[31] China's slowing economy and rising labor costs have also encouraged greater Chinese overseas FDI in order to help firms diversify risk and expand business opportunities beyond

[31] The composition of Chinese FDI sectors has changed over the past few years. For example, according to AEI/Heritage Foundation, in 2010, 67% of Chinese FDI outflows were in energy and metals sectors, but by 2015, this level dropped to 29%, caused in part by large levels of Chinese FDI in transportation, finance, real estate, and technology sectors.

the China market, and, in some cases, to relocate less competitive firms from China to low-cost countries. China's Ministry of Foreign Trade (MOFCOM) reports that in 2018, Chinese nonfinancial FDI in BRI countries totaled $15.6 billion, up 8.9% over the previous year.[32] Additionally, increased FDI outflows may be the result of the Chinese government attempting to diversify its foreign exchange reserve holdings (which totaled $3.1 trillion as of April 2019—by far the world's largest holder). The largest foreign investors in China (based on FDI stock through 2017) were Hong Kong (52.6% of total),[33] the British Virgin Islands (10.6%), Japan (6.1%), Singapore (4.0%), and Germany (3.2%) (see Table 2).

Table 2. Chinese data on top ten sources of China's FDI inflows to China: 1979-2017 ($ billions and percentage of total)

Country	**Estimated Cumulative Utilized FDI: 1979-2017**	
	Amount	% of Total
Total	2,688	100
Hong Kong	1,241	46.2
British Virgin Islands	286	10.6
Japan	165	6.1
Singapore	108	4.0
Germany	87	3.2
S. Korea	73	2.7
U.S.	72	2.7
Cayman Islands	49	1.8
The Netherlands	37	1.4
Taiwan	33	1.2

Source: IMF Coordinated Direct Investment Survey

[32] Xinhuanet, "China's ODI sees stable development in 2018," January 16, 2019, available at http://www.xinhuanet.com/english/2019-01/16/c_137749000.htm

[33] Much of the FDI originating from Hong Kong may originate from other foreign investors, such as Taiwan. In addition, some Chinese investors might be using these locations to shift funds overseas in order to re-invest in China to take advantage of preferential investment policies (this practice is often referred to as "round-tipping"). Thus, the actual level of FDI in China may be overstated.

Factors Driving China's FDI Outflow Strategy

A key aspect of China's economic modernization and growth strategy during the 1980s and 1990s was to attract FDI into China to help boost the development of domestic firms. Investment by Chinese firms abroad was sharply restricted. However, in 2000, China's leaders initiated a new "go global" strategy, which sought to encourage Chinese firms (primarily SOEs) to invest overseas. One key factor driving this investment is China's massive accumulation of foreign exchange reserves. Traditionally, a significant level of those reserves has been invested in relatively safe but low-yielding assets, such as U.S. Treasury securities. On September 29, 2007, the Chinese government officially launched the China Investment Corporation (CIC) in an effort to seek more profitable returns on its foreign exchange reserves and diversify away from its U.S. dollar holdings.[34] The CIC was originally funded at $200 billion, making it one of the world's largest sovereign wealth funds.[35] Another factor behind the government's drive to encourage more outward FDI flows has been to obtain natural resources, such as oil and minerals, deemed by the government as necessary to sustain China's rapid economic growth.[36] Finally, the Chinese government has indicated its goal of developing globally competitive Chinese firms with their own brands. Investing in foreign firms, or acquiring them, is viewed as a method for Chinese firms to obtain technology, management skills, and often, internationally recognized brands, needed to help Chinese firms become more globally competitive. For example, in April 2005, Lenovo Group Limited, a Chinese computer company, purchased IBM Corporation's personal computer division for $1.75 billion.[37] The largest destinations of cumulative Chinese FDI outflows through 2017 were Hong Kong (54.2% of total), the Cayman Islands (13.9%), the British Virgin Islands (6.7%), and the United States (3.7%) (see Table 3).

---

[34] See CRS Report RL34337, *China's Sovereign Wealth Fund*, by Michael F. Martin.
[35] At the end of 2015, CIC's assets totaled $810 billion.
[36] Chinese oil and mineral companies are dominated by SOEs.
[37] The Chinese government is believed to be Lenovo's largest shareholder.

Table 3. Major destinations of Chinese nonfinancial FDI outflows by stock through 2017 ($ billions and percent of total)

Destination	Stock of FDI through 2017	Share of FDI Stock through 2017 (%)
Total	1,809	—
Hong Kong	981	54.2
Cayman Islands	251	13.9
British Virgin Islands	122	6.7
United States	67	3.7
Singapore	45	2.5
Australia	36	2.0
United Kingdom	20	1.1

Source: China Natural Bureau of Statistics.

Note: Ranked according to the top seven destinations of the stock of Chinese FDI outflows through 2017.

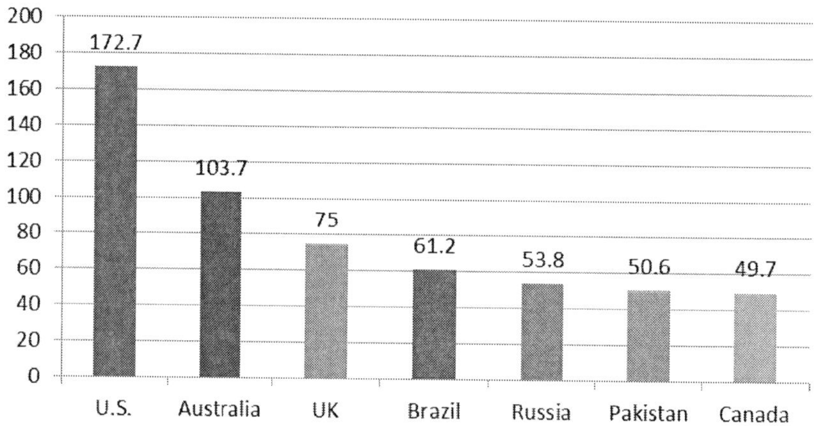

Source: AEI China Global Investment Tracker.

Notes: Data includes outward FDI and valued at $100 million or more. These data differ significantly from official U.S. and Chinese government data which rely on different methodologies.

Figure 14. AEI Estimates of Chinese Cumulative Outward Investment by Major Destination: 2005-2017 ($ billions).

A significant level of Chinese FDI that goes to Hong Kong, the British Virgin Islands, and the Cayman Islands likely is redirected elsewhere. The American Enterprise Institute (AEI) maintain the China Global Investment Tracker (CGIT), a database that has been developed to track the actual flows (from the parent company to the final destination) of Chinese investment globally. The CGIT database tracks FDI valued at $100 million or more (which it refers to as "China's outward non-bond investment").[38] These data differ significantly from official Chinese FDI outflow data. The CGIT data on the top destinations of total Chinese outward non-bond outward investment from 2005 to 2017 included the United States ($172.7 billion), Australia ($103.7 billion), the United Kingdom ($75 billion), Brazil ($61.2 billion), and Russia ($53.8) (see Figure 14).[39]

China's Merchandise Trade Patterns

Economic reforms and trade and investment liberalization have helped transform China into a major trading power. Chinese merchandise exports rose from $14 billion in 1979 to $2.5 trillion in 2018, while merchandise imports grew from $18 billion to $2.1 trillion (see Table 4 and Figure 15). China's rapidly growing trade flows have made it an increasingly important (and often the largest) trading partner for many countries. According to China, it was the largest trading partner for 130 countries in 2013.[40] From 2000 to 2008, the annual growth of China's merchandise exports and imports averaged 25.1% and 24.2%, respectively.

However, China's exports and imports fell by 15.9% and 11.2%, respectively, due to the impact of the global financial crisis. China's trade recovered in 2010 and 2011, with export growth averaging 25.8% and import growth averaging 31.9%. However, since that time, China's trade growth slowed sharply. From 2012 to 2014, China's exports and imports grew at an average annual rate of 7.2% and 4.1%, respectively.

---

38 AEI/Heritage Foundation's methodology do not use the standard measurement of FDI, which generally includes foreign ownership or control of at least 10% share or control of an entity.

39 The CGIT also estimates the flow of Chinese FDI to the United States in 2017 at $24.5 billion (compared to $54.6 billion in 2016), making the United States the largest destination of Chinese outward FDI. China's largest U.S. acquisition in 2017 was HNA's purchase of CIT Group's aircraft leasing business for $10.4 billion.

40 *China.org.cn*, "Promoting China-Japan relations through Culture," June 18, 2014, at http://www.china.org.cn/ opinion/2014-06/18/content_32690843.htm.

From 2015 to 2016 exports and imports fell by an average rate of 4.7% and 11.6%, respectively (see Figure 16), reflecting a sluggish global economy and a decline in commodity prices (such as oil and ores).

Table 4. China's global merchandise trade: 1979-2018 ($ billions)

Year	Exports	Imports	Trade Balance
1979	13.7	15.7	-2.0
1980	18.1	19.5	-1.4
1985	27.3	42.5	-15.3
1990	62.9	53.9	9.0
1995	148.8	132.1	16.7
2000	249.2	225.1	24.1
2001	266.2	243.6	22.6
2002	325.6	295.2	30.4
2003	438.4	412.8	25.6
2004	593.4	561.4	32.0
2005	762.0	660.1	101.9
2006	969.1	791.5	177.6
2007	1,218.0	955.8	262.2
2008	1,428.9	1,131.5	297.4
2009	1,202.0	1,003.9	198.2
2010	1,578.4	1,393.9	184.5
2011	1,899.3	1,741.4	157.9
2012	2,050.1	1,817.3	232.8
2013	2,210.7	1,949.3	261.4
2014	2,343.2	1,963.1	380.1
2015	2,280.5	1,601.8	678.8
2016	2,135.3	1,524.7	610.6
2017	2,279.2	1,790.0	489.2
2018	2,491.4	2,109.0	382.4

Source: Global Trade Atlas and China's Customs Administration.

However, in 2017, China's exports and imports rose by 6.7% and 17.4%, respectively. Exports and imports in 2018 rose by 9.3 and 17.8%, respectively. However, during the first three months of 2001, China's exports grew by 1.0%, while imports fell 1.1% year-over-year.

China's merchandise trade surplus grew sharply from 2004 to 2008, rising from $32 billion to $297 billion. That surplus fell each year over the next three years, dropping to $158 billion in 2011. However, it rose in each of the next four years, reaching a record $679 billion in 2015 before falling to $611 billion in 2016, $489 billion in 2017, and $382 billion in 2018. In 2009, China overtook Germany to become both the world's largest merchandise exporter and the second-largest merchandise importer (after the United States). In 2012, China overtook the United States as the world's largest merchandise trading economy (exports plus imports). As indicated in Figure 17, China's share of global merchandise exports grew from 2.0% in 1990 to 14.1% in 2015, but fell to 13.4% in 2016 and to 13.2% in 2017.

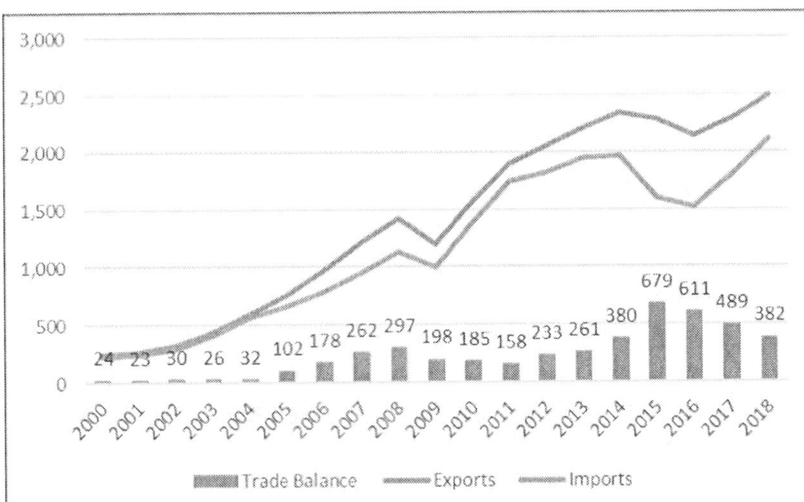

Source: World Trade Atlas and China's Customs Administration.

Note: Data are in U.S. dollars which may be impacted by changes in exchange rates.

Figure 15. China's Merchandise Trade: 2000-2018 ($ billions).

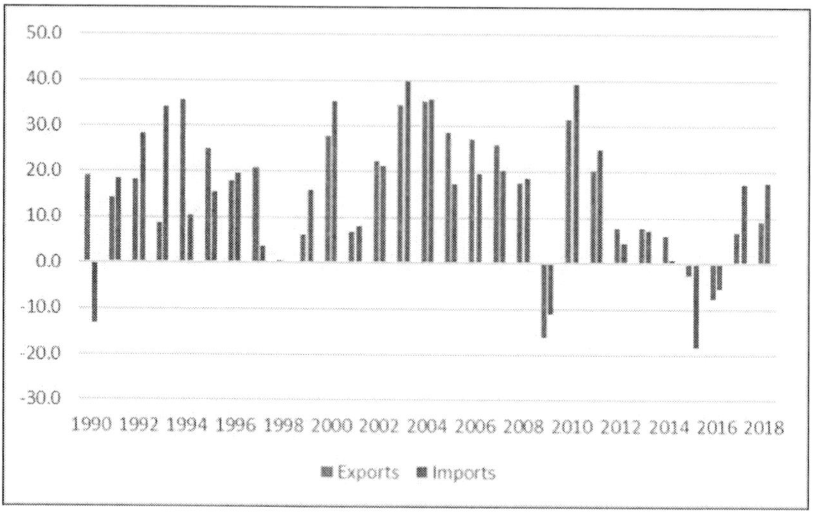

Source: Global Trade Atlas and China's Customs Administration.

Note: Data are in U.S. dollars which may be impacted by changes in exchange rates as well as commodity prices.

Figure 16. Annual Change in China's Merchandise Exports and Imports: 2000-2018 (percentage).

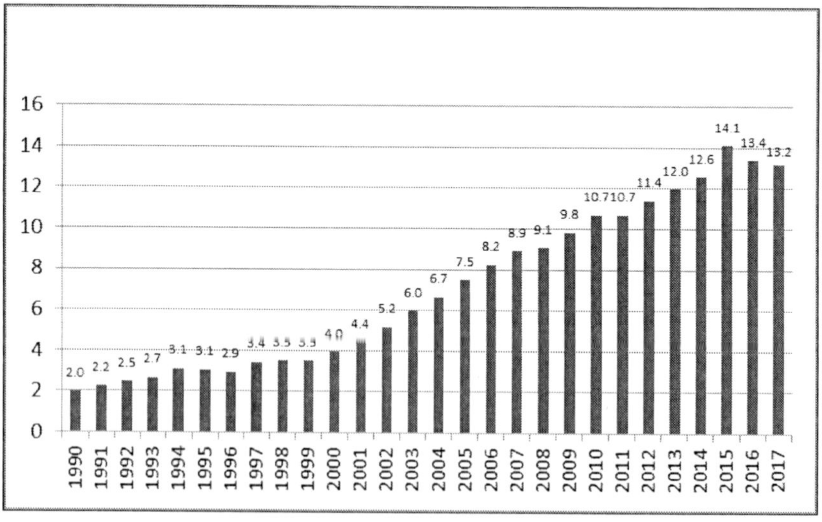

Source: Economist Intelligence Unit. Note: Data for 2017 are estimated.

Figure 17. China's Share of Global Merchandise Exports: 1990-2017 (percentage).

China's Major Trading Partners

Table 5. China's major merchandise trading partners in 2018 ($ billions)

Country	Total Trade	Chinese Exports	Chinese Imports	China's Trade Balance
European Union	681	408	273	135
United States	631	477	154	323
ASEAN	575	318	257	61
Japan	327	147	180	-33
South Korea	313	109	204	-95
Hong Kong	310	302	8	294
Taiwan	225	48	177	-129

Source: China's Customs Administration.

Notes: Rankings according to China's total trade in 2018. China's bilateral trade data often differ from that of its trading partners.

Table 5 lists official Chinese trade data on its seven largest trading partners in 2018 (based on total trade). These include the 28 countries that make up the European Union (EU28), the United States, the 10 nations that constitute the Association of Southeast Asian Nations (ASEAN), Japan, South Korea, Hong Kong, and Taiwan.[41] China's top three export markets were the United States, the EU28, ASEAN,[42] while its top sources for imports were the EU28, ASEAN, and South Korea. According to Chinese data, it maintained large trade surpluses with the United States ($282 billion), Hong Kong ($274 billion) and the EU28 ($129 billion), and reported large trade imbalances with Taiwan ($112 billion) and South Korea ($74 billion). China's trade data differ significantly from those of many of its trading partners. These differences appear to be largely caused by how China's trade via Hong Kong is counted in official Chinese trade data. China treats a large share of its exports through Hong Kong as Chinese exports to Hong Kong for statistical purposes, while many countries that import Chinese products through Hong Kong generally

---

[41] ASEAN members include Brunei, Cambodia, Indonesia, Laos, Malaysia, Myanmar (Burma), the Philippines, Singapore, Thailand, and Vietnam.

[42] Much of Chinese exports to Hong Kong are later re-shipped elsewhere.

attribute their origin to China for statistical purposes, including the United States.[43]

Major Chinese Trade Commodities

Table 6. Major Chinese Merchandise Imports in 2018

HS Code	Description	$ Billions	Percentage of Total Exports	2018/2017 % Change
Total Commodities	2,117	100	18.3	
85	Electrical machinery and equipment	522	24.6	14.2
27	Mineral fuel, oil etc.	348	16.4	41.5
84	Nuclear reactors, boilers, and machinery	202	9.6	18.8
26	Ores, slag, and ash	135	6.4	8.3
90	Optical, photographic, cinematographic, measuring checking, precision, medical or surgical instruments and apparatus; parts and accessories thereof	103	4.8	5.2
87	Vehicles, (except railway, and parts trucks, and bicycles)	81	3.9	2.9
71	Precious stones and metals	61	2.9	328.6
74	Copper and articles thereof	48	2.3	15.5
12	Oil seeds, misc. grain, plants, and fruit	43	2.1	-2.7
30	Pharmaceutical products	28	1.3	10.8

Source: *World Trade Atlas*, using official Chinese statistics.

Note: Top 10 imports in 2018, two-digit level, harmonized tariff system.

[43] See CRS Report RS22640, *What's the Difference?—Comparing U.S. and Chinese Trade Data*, by Michael F. Martin.

China's abundance of low-cost labor has made it internationally competitive in many low-cost, labor-intensive manufactures. As a result, manufactured products constitute a significant share of China's trade. A substantial amount of China's imports is comprised of parts and components that are assembled into finished products, such as consumer electronic products and computers, and then exported. Often, the value-added to such products in China by Chinese workers is relatively small compared to the total value of the product when it is shipped abroad.

Table 7. Major Chinese merchandise exports in 2018

HS Code	**Description**	**$ Billions**	**Percentage of Total Exports**	**2017/2016 % Change**
Total Commodities	2,489	100	9.2	
85	Electrical machinery	664	26.7	10.6
84	Nuclear reactors, boilers, and machinery	429	17.3	12.0
94	Furniture and bedding	96	3.9	5.6
39	Plastics and articles thereof	80	3.2	12.1
87	Vehicles, except railway, and parts	75	3.0	11.4
61	Apparel articles and accessories, knit or crochet	74	3.0	1.9
62	Apparel articles and accessories, woven	71	2.9	-3.3
90	Optical, photographic, cinematographic, measuring checking, precision, medical or surgical instruments and apparatus; parts and accessories thereof	71	2.9	0.8
73	Articles of iron or steel	65	2.6	13.0
29	Organic chemicals	60	2.4	20.3

Source: *World Trade Atlas*, using official Chinese statistics.

Note: Top 10 exports in 2018, two-digit level, harmonized tariff system.

China's top 10 imports and exports in 2018 are listed in Table 6 and Table 7, respectively, using the harmonized tariff system (HTS) on a two-digit level. Major imports included electrical machinery and equipment;[44] mineral fuels; nuclear reactors, boilers, and machinery (such as automatic data process machines and machines to make semiconductors); ores; and optical, photographic, medical, or surgical instruments. China's biggest exports were electrical machinery and equipment; nuclear reactors, boilers, and machinery; furniture; plastics; and vehicles.

Major Long-Term Challenges Facing the Chinese Economy

China is currently undergoing a major restructuring of its economic model. Policies that were employed in the past to essentially produce rapid economic growth at any cost were very successful. However, such policies have entailed a number of costs (such as heavy pollution, widening income inequality, overcapacity in many industries, an inefficient financial system, rising corporate debt, and numerous imbalances in the economy) and therefore the old growth model is viewed by many economists as no longer sustainable. China has sought to develop a new growth model ("the new normal") that promotes more sustainable (and less costly) economic growth that puts greater emphasis on private consumption and innovation as the new drivers of the Chinese economy. Implementing a new growth model that sustains healthy economic growth could prove challenging unless China is able to effectively implement new economic reforms. Many analysts warn that without such reforms, China could face a period of stagnant economic growth and living standards, a condition referred to by economists as the "middle-income trap" (Several of these challenges are discussed below.

[44] This includes electrical machinery and equipment and parts thereof; sound recorders and reproducers, television image and sound recorders and reproducers, and parts and accessories of such articles.

China's Incomplete Transition to a Market Economy

Despite China's three-decade history of widespread economic reforms, Chinese officials contend that China is a "socialist-market economy." This appears to indicate that the government accepts and allows the use of free market forces in a number of areas to help grow the economy, but the government still plays a major role in the country's economic development.

Industrial Policies and SOEs

According to the World Bank, "China has become one of the world's most active users of industrial policies and administrations."[45] China's State Council has said that there are currently 150,000 SOEs at the central and local government level.[46] China's SOEs may account for up of 50% of non-agriculture GDP.[47] In addition, although the number of SOEs has declined sharply, they continue to dominate a number of sectors (such as petroleum and mining, telecommunications, utilities, transportation, and various industrial sectors); are shielded from competition; are the main sectors encouraged to invest overseas; and dominate the listings on China's stock indexes.[48] One study found that SOEs constituted 50% of the 500 largest manufacturing companies in China and 61% of the top 500 service sector enterprises.[49] Not only are SOEs dominant players in China's economy, many are quite large by global standards.

---

[45] The World Bank, *China: 2030*, February 27, 2012, p. 114.

[46] The State Council of the People's Republic of China, "Why is China reforming State-owned enterprises?," February 16, 2017, available at http://english.gov.cn/news/video/2017/02/16/content_281475569025065.htm.

[47] U.S.-China Economic and Security Review Commission, An Analysis of State-owned Enterprises and State Capitalism in China, by Andrew Szamosszegi and Cole Kyle, October 26, 2011, p. 1.

[48] The nature of China's SOEs has become increasing complex. Many SOEs appear to be run like private companies. For example, and a number of SOEs have made initial public offerings in China's stock markets and those in other countries (including the United States), although the Chinese government is usually the largest shareholder. It is not clear to what extent the Chinese government attempts to influence decisions made by the SOE's which have become shareholding companies.

[49] Xiao Geng, Xiuke Yang, and Anna Janus, *State-owned Enterprises in China, Reform Dynamics and Impacts*, 2009, p. 155.

China and Global Steel Overcapacity

China has become a major global producer of steel. From 2001 to 2016, China's production of raw steel rose from 152 million metric tons to 805 million metric tons, an increase of 459.9%. During this period, China's share of global production rose from 17.9% to 50.3% and China accounted for 87.1% of the increase in global steel production.[50] While much of China's increased steel capacity has been in response to domestic demand (resulting from China's large-scale fixed investment), it is also a major exporter. In 2015, China was the second-largest exporter of semifinished and steel products (after the EU) at 111.6 million metric tons, or 24.1% of global total. Concerns have been raised over the past few years over the effects of increased global steel production despite falling global steel demand. Following an EU Symposium on Excess Capacity and Structural Adjustment in the Steel Sector in April 2016, then-U.S. Secretary of Commerce Penny Pritzker and then-USTR Michael Froman issued a joint statement that said that the "U.S. steel industry is in a crisis driven in large part by global excess capacity.... led by unsustainable expansion in steelmaking capacity by China," and noted that global steel overcapacity had impacted the U.S. industry through price declines, decreased profitability, and over 13,000 jobs lost.[51] Many analysts contend that China's steel industry is heavily supported by government entities at the central and local government level through low-cost credit and subsidies. These enable many Chinese steel firms to operate, even when they are unprofitable and heavily in debt—these are termed by some as "zombies." The government is reluctant to allow such firms to go bankrupt due to concerns over effects layoffs could have on social stability. One study by Renmin University estimated that half of Chinese steel firms were "zombie enterprises."[52] In April 2016, the USTR released a list of 86 government subsidies from 2011 to 2014 given to Hebei Iron & Steel Company, one of the largest steel companies in China, which was obtained from the company's annual reports.[53]

In February 2016, the Chinese government announced plans to shut 100 million to 150 million metric tons of crude steel capacity over the next five years and cut 500,000 jobs.[54] The USTR's 2016 report on China's WTO compliance stated that China had committed to "ensure that no central government plans, policies, directives, guidelines, lending or subsidization targets the net expansion of steel capacity," and to take steps to reduce existing capacity, including "actively and appropriately dispose of 'zombie enterprises' through bankruptcies and other means."[55]

---

[50] Data from World Steel Association, at https://www.worldsteel.org/steel-by-topic/statistics/Steel-StatisticalYearbook-.html.

[51] USTR, *Press Release*, April 16, 2016, available at https://ustr.gov/about-us/policy-offices/press-office/press-releases/ 2016/april/statement-secretary-pritzker-and-ustr#.

[52] The Wall Street Journal, *China Steels Its Resolve, But 'Zombies' Abound,* July 29, 2016, available at http://blogs.wsj.com/chinarealtime/2016/07/29/csteel0729/.

[53] Inside U.S. Trade, China Trade Extra, *U.S. Lays Out Subsidies To Chinese Steel Firm In 'Room Document' Circulated At WTO Meeting*, April 29, 2016.

[54] Bloomberg News, *China Expects 1.8 Million Coal, Steel Layoffs on Capacity Cuts*, February 29, 2016, available at https://www.bloomberg.com/news/articles/2016-02-29/china-expects-1-8-million-coal-steel-layoffs-on-capacity-cuts.

[55] USTR, *2016 Report to Congress On China's WTO Compliance*, January 2017, p. 6.

Fortune's 2016 list of the world's 500 largest companies includes 103 Chinese firms (compared to 29 listed firms in 2007).[56] Of the 103 Chinese firms listed, Fortune identified 75 companies (73% of total) where the government owned 50% or more of the company. Together, these 75 firms in 2016 generated $7.2 trillion in revenues, had assets valued at $20.7 trillion, and employed 16.2 million workers. Of the 28 other Chinese firms on the Fortune 500 list, several appear to have financial links to the Chinese government.

A State-Dominated Banking Sector, Excess Credit, and Growing Debt

China's banking system is largely dominated by state-owned or state-controlled banks. According to one analyst, the mangers of China's state banks are drawn from the ranks of the Chinese Communist Party cadre system, which "enables the party and government leaderships to exert influence over bank lending."[57] In 2015, the top five largest banks in China in terms of assets were state-owned entities.[58] The percentage share of assets held by state-owned commercial banks (including the five large state-owned banks), the three government policy banks,[59] and joint-stock commercial banks (where government entities are a major stock holder), together accounted for 68.5% of total bank assets in China.[60] Foreign participation in China's banking system is relatively small, accounting for 1.6% of total bank assets.[61] SOEs are believed to receive preferential credit treatment by government banks, while private firms must often pay

---

[56] The listing can be found at http://beta.fortune.com/global500/.

[57] Mercator Institute for China Studies, *China's Political System*, 2017, p. 213.

[58] These were the Industrial & Commercial Bank of China, China Construction Bank, Agricultural Bank of China, Bank of China, and Bank of Communications.

[59] These include Export Import Bank of China, Agricultural Development Bank of China, and China Development Bank.

[60] Bejkovsky, Ing. Jan, *State Capitalism in China: The Case of the Banking Sector*, August 2016, available at http://globalbizresearch.org/IAR16_Vietnam_Conference_2016_Aug/docs/doc/PDF/VS611.pdf.

[61] Ibid.

higher interest rates or obtain credit elsewhere. According to one estimate, SOEs accounted for 85% ($1.4 trillion) of all bank loans in 2009.[62] It is believed that oftentimes SOEs do not repay their loans, which may have saddled the banks with an ever-increasing amount of nonperforming loans.[63] Many analysts contend that one of the biggest weaknesses of the banking system is that it lacks the ability to ration and allocate credit according to market principles, such as risk assessment.

The Chinese central government uses the banking system to boost credit in order to help meet its GDP growth objectives and to, when needed, offset the impact of global economic downturns, such as after the 9/11 terrorist attacks and the global financial crisis. From 2007 to 2016, China's domestic credit increased in dollar terms by 218% (see Figure 18), and as a share of GDP, the level rose from 125% to 212%. As indicated in Figure 19, China's combined household, corporate, and government debt levels as a percentage of GDP as of mid-2016 are comparable to those of the United States and South Korea and lower than those of Japan and the European Union. However, China's debt levels (in both dollars and as a percentage of GDP) have risen sharply within a relatively short time, which, some have speculated, could spark an economic crisis in China in the future. From 2006 year-end to mid-2016, China's total nonfinancial sector debt as a percentage of GDP increased from 143% to 254% (up 111 percentage points). Much of the rise in that debt came from the corporate sector, which, as a percentage of GDP, rose from 107% in 2006 to 171% in mid-2016 (up 64 percentage points). In dollar terms, China's corporate debt rose from $3 trillion to $17.8 trillion (up $14.8 trillion) and currently greatly exceeds U.S. corporate debt levels (see Figure 20).

---

[62] The Economist, *State Capitalism's Global Reach, New Masters of the Universe, How State Enterprise is Spreading*, January 21, 2012.

[63] According to Chinese officials, the ratio of nonperforming in China of commercial banks at the end of 2016 was 1.7% ($220 billion), although some analysts contend that this figure could be much higher. See, Reuters, "Chinese commercial banks' NPL ratio at 1.74 pct – regulator," January 25, 2017, available at http://www.reuters.com/article/ china-banks-npl-idUSB9N1FA01A.

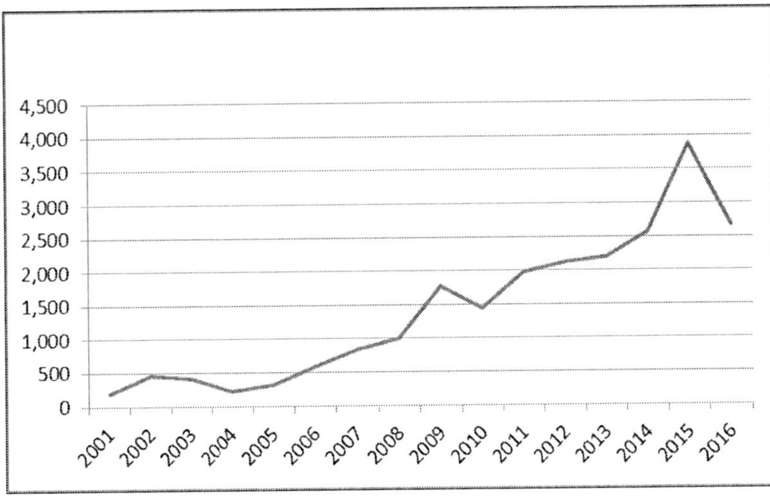

Source: Economist Intelligence Unit.

Figure 18. Annual Change in the Stock of China's Domestic Credit 2001-2016 ($ billions).

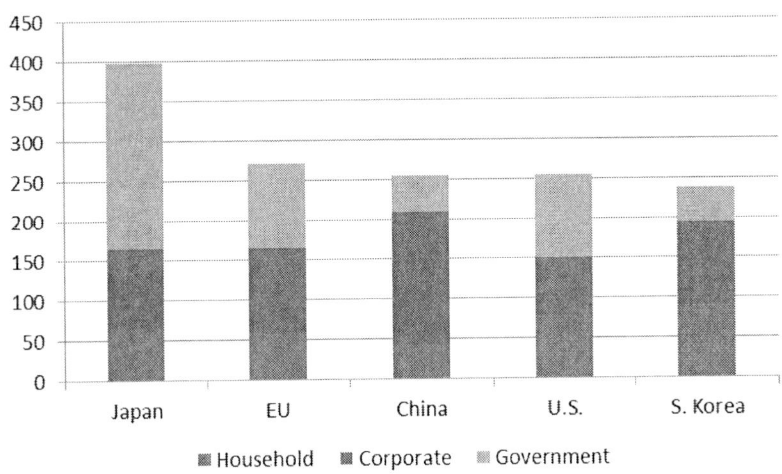

Source: Bank for International Settlements.

Note: * As of second quarter 2016.

Figure 19. Core Debt of Nonfinancial Sectors in 2016* as a Percentage of GDP for Selected Economies (percentage).

Several observers have warned that China's credit growth may be too extensive and could undermine future growth by sharply boosting debt levels, causing overcapacity in many industrials (especially extending credit to firms that are unprofitable to keep them operating), contributing to bubbles (such as in real estate), and reducing productivity by proving preferential treatment to SOEs and other government-supported entities.

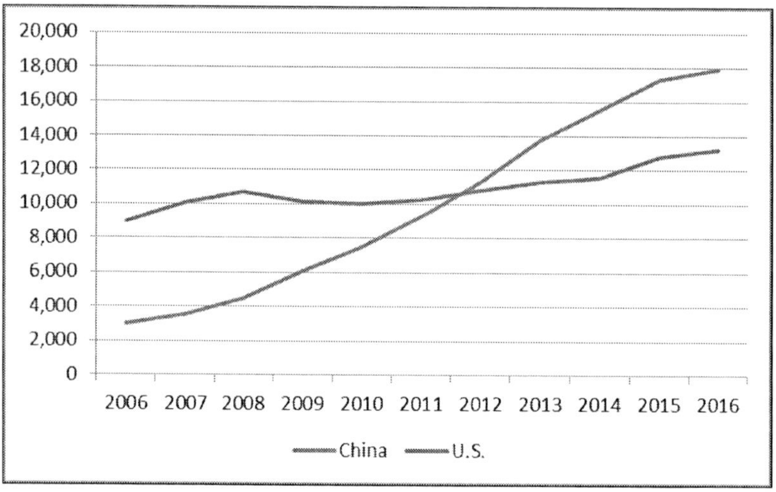

Source: Bank for International Settlements.
Note: *As of second quarter 2016.

Figure 20. U.S. and Chinese Corporate Debt: 2006-2016* ($ billions).

Local government debt is viewed as a big problem in China, largely because of the potential impact it could have on the Chinese banking system. During the beginning of the global financial slowdown, many Chinese subnational government entities borrowed extensively to help stimulate local economies, especially by supporting infrastructure projects. In December 2013, the Chinese National Audit Office reported that from the end of 2010 to mid-year 2013, local government debt had increased by 67% to nearly $3 trillion.[64] The Chinese government reported that local government debt rose to $4.3 trillion as of 2015. Efforts have been made

[64] The Wall Street Journal, *Xi Faces Test over China's Local Debt; Risks From Debt are Still Controllable, Audit Office Says*, December 30, 2013.

over the past few years by the central government to restructure local government debt and restrict local government borrowing, with mixed success, according to some press reports, because of pressures on local governments to maintain rapid economic growth.[65]

Many economists blame China's closed capital account for much of China's debt problems. The Chinese government has maintained restrictions on capital inflows and outflows for many years, in part to control the exchange of its currency, the renminbi (RMB), against the dollar and other currencies in order to boost exports. Many argue the Chinese government's restrictions on capital flows have greatly distorted financial markets in China, preventing the most efficient use of capital, such as overinvestment in some sectors (such as real estate) and underinvestment in others (such as services).

Environmental Challenges

China's economic growth model has emphasized the growth of heavy industry in China, much of which is energy-intensive and high polluting. The level of pollution in China continues to worsen, posing serious health risks to the population. The Chinese government often disregards its own environmental laws in order to promote rapid economic growth. China's environmental challenges are illustrated by the following incidents and reports.

- A 2018 report by ExxonMobil estimated that China contributed about 60% of the growth in global CO2 emissions from 2000 to 2016, and that its emissions would surpass the combined CO2 levels of the United States and EU by 2025.[66]

---

[65] See for example, the Financial Times, "China local governments revive off-budget fiscal stimulus," September 21, 2016, available athttps://www.ft.com/content/b303f280-7f14-11e6-8e50-8ec15fb462f4.

[66] ExxonMobil, *2018 Outlook for Energy, A View to 2040*, 2018, p. 60, available at http://cdn.exxonmobil.com/~/media/global/files/outlook-for-energy/2018/2018-outlook-for-energy.pdf.

- A 2017 OECD report estimated the health costs of China's air pollution in 2015 at $1.4 trillion, equivalent to 7.8% of its GDP.[67]
- A 2015 study by the Rand Corporation estimated that the costs (in terms of health impact and lost productivity) from China's air pollution were equal to 6.5% of GDP each year from 2000 to 2010. It further estimated the costs as a percentage of GDP of water pollution and soil degradation at an additional 2.1% and 1.1%, respectively.[68]
- On August 12, 2015, a series of large explosions in several warehouses containing chemicals occurred in the Chinese port city of Tianjin, claiming the lives of at least 163 people. Some press reports have blamed poor government enforcement of environmental regulations for the disaster. For example, some in China have questioned why dangerous chemicals were warehoused so close to residential areas and have raised concerns over the extent of chemical contamination in the area that may have resulted from the explosions.
- The U.S. Embassy in Beijing, which monitors and reports air quality in China based on an air quality index of particulate matter (developed by the U.S. Environmental Protection Agency) considered to pose a health concern, reported that the air quality in Beijing for a majority of the days in January 2013 ranged from "unhealthy" to "hazardous" (based on 24-hour exposure) and, on a few days, it recorded high readings that were "beyond index."[69] The level of poor air quality in Beijing was termed by some in China as

[67] OECD, *The Rising Cost of Ambient Air Pollution thus far in the 21st Century, Results from the BRIICS and the OECD Countries,* July 2017, p. 22, available at http://www.oecd-ilibrary.org/docserver/download/d1b2b844-en.pdf?expires=1517681542&id=id&accname=guest&checksum=9B43144FCF78931DCE50EBEC9B8F84E8.

[68] The Rand Corporation, *Cost of Selected Policies to Address Air Pollution in China*, 2015, p. 3, available at http://www.rand.org/content/dam/rand/pubs/research_reports/RR800/RR861/RAND_RR861.pdf.

[69] Hazardous is the worst category for air quality used by the U.S. embassy, based on a numerical value of its index ranging from 301 to 500. A measurement of below 50 is considered good. On several occasions, the air quality index in Beijing has surpassed 500, and on January 12, 2013, it reportedly hit 755.

"Airpocalypse," and reportedly forced the government to shut down some factories and reduce the level of official cars on the road.[70] On December 9, 2013, China's media reported that half of China was blanketed by smog.[71] The U.S. Consulate General in Shanghai reported that were a number of days in December 2013 where its measurement of the air quality in Shanghai was hazardous or very unhealthy, and during some time periods on December 5, 2013, its readings were "beyond index." According to the U.S. Embassy in Beijing, from 2008 to 2015, nearly two-thirds of the days in Beijing had air pollution considered to be unhealthy.[72]

- In February 2013, China's Geological Survey reportedly estimated that 90% of all Chinese cities had polluted groundwater, with two-thirds having "severely polluted" water.[73]
- According to a 2012 report by the Asian Development Bank, less than 1% of the 500 largest cities in China meet the air quality standards recommended by the World Health Organization, and 7 of these are ranked among the 10 most polluted cities in the world.[74]

The Chinese government has indicated that it is taking steps to reduce energy consumption, boost enforcement of environmental laws and regulations, reduce coal usage by expanding the use of cleaner fuels (such as natural gas) to general power, and relocate high-polluting factories away from large urban areas, although such efforts have had mixed results on the overall level of pollution in China.[75] In addition, China has become a major

---

[70] National Public Radio, "Beijing's 'Airpocalypse' Spurs Pollution Controls, Public Pressure," January 14, 2013.

[71] *Xinhua*, December 9, 2013.

[72] The BBC, "China pollution: First ever red alert in effect in Beijing," December 8, 2015, available at http://www.bbc.com/news/world-asia-china-35026363.

[73] *New York Times*, "Concerns Grow About 'Severely Polluted' Water in China's Cities," February 20, 2013.

[74] The Asian Development Bank, *Toward an Environmentally Sustainable Future, Country Environmental Analysis of the People's Republic of China*, 2012, p. xviii.

[75] Bloomberg, "China Is Winning Its War on Air Pollution, at Least in Beijing," January 11, 2018, available at https://www.bloomberg.com/news/articles/2018-01-11/china-is-winning-its-war-on-air-pollution-at-least-in-beijing.

global producer and user of clean and renewable energy technology. In January 2017, the Chinese government said it would spend $361 billion on renewable energy power generation by 2020.[76]

Corruption and the Relative Lack of the Rule of Law

The relative lack of the rule of law in China has led to widespread government corruption, financial speculation, and misallocation of investment funds. In many cases, government "connections," not market forces, are the main determinant of successful firms in China. Many U.S. firms find it difficult to do business in China because rules and regulations are generally not consistent or transparent, contracts are not easily enforced, and intellectual property rights are not protected (due to the lack of an independent judicial system). The relative lack of the rule of law and widespread government corruption in China limit competition and undermine the efficient allocation of goods and services in the economy. A *New York Times* article reported that (former) Chinese Premier Wen Jiabao's family controlled assets worth at least $2.7 billion.[77] One study estimates that between 2001 and 2010, China was the world's largest source of illicit capital outflows at $3.8 trillion.[78] A 2012 survey by the Pew Research Center's Global Attitudes Project reported that 50% of respondents said that corrupt officials are a very big problem (up from 39% in 2008).[79] Chinese officials often identify government corruption as the greatest threat to the Chinese Communist Party and the state. The Chinese government's anticorruption watchdog reported that 106,000 officials were found guilty of corruption in 2009.[80] Since assuming power in 2012, Chinese Xi Jinping has carried out an extensive anticorruption drive. China has reportedly sought cooperation with the United States to obtain

[76] Reuters, "China to plow $361 billion into renewable fuel by 2020," January 4, 2017, at https://www.reuters.com/ article/us-china-energy-renewables/china-to-plow-361-billion-into-renewable-fuel-by-2020-idUSKBN14P06P.

[77] *New York Times*, "Billions in Hidden Riches for Family of Chinese Leader," October 25, 2012.

[78] Global Financial Integrity, *Chinese Economy Lost $3.79 Trillion in Illicit Financial Outflows Since 2000, Reveals New GFI Report,* October 25, 2012. It is not known how much of the illicit financial outflows in China are directly linked to government corruption.

[79] Pew Research Global Attitudes Project, *Growing Concerns in China about Inequality, Corruption*, October 16, 2012.

[80] BBC News, "Corruption Up Among China Government Officials," January 8, 2010.

extradition of 150 alleged corrupt officials who have fled to the United States.[81] However, many analysts contend that government anticorruption campaigns are mainly used to settle political scores with out-of-favor officials. Some analysts contend that President's Xi anticorruption drive is more about consolidating his own political than instituting reforms.[82] In addition, there are some indicators that the current anticorruption campaign may be having a negative impact on the Chinese economy, due to hesitancy by some local officials to pursue projects they feel will lead to scrutiny from the central government.[83] Many observers argue that meaningful progress against government corruption cannot occur without greater government transparency, a system of checks and balances, a free press, Internet freedom, and an independent judiciary.[84] In October 2014, China held its fourth Plenum of the 18th Party Conference. The meeting focused on the need to enhance the rule of law in China, but emphasized the leading role of the Communist Party in the legal system.[85]

China maintains a weak and relatively decentralized government structure to regulate economic activity in China. Laws and regulations often go unenforced or are ignored by local government officials. As a result, many firms cut corners in order to maximize profits. This has led to a proliferation of unsafe food and consumer products being sold in China or exported abroad. Lack of government enforcement of food safety laws led to a massive recall of melamine-tainted infant milk formula that reportedly killed at least four children and sickened 53,000 others in 2008.

---

81 The Atlantic Monthly, *How to Discipline 90 Million People*, April 2015, available at http://www.theatlantic.com/ international/archive/2015/04/xi-jinping-china-corruption-political-culture/389787/.

82 The Washington Post, *China's Leader, Xi Jinping, Consolidates Power with Crackdowns on Corruption, Internet*, October 3, 2013.

83 A study reportedly issue by Bank of America Merrill Lynch estimated that China's current antigraft campaign would cost the Chinese economy more than $100 billion. See, BBC, China Blog, *the Real Costs of China's Anti-Corruption Crackdown*, April 2, 2014, available at http://www.bbc.com/news/blogs-china-blog-26864134.

84 *New York Times*, "Chinese Officials Find Misbehavior Now Carries Cost," December 25, 2012.

85 The Diplomat, "4 Things We Learned from China's 4th Plenum," October 23, 2014, available at http://thediplomat.com/2014/10/4-things-we-learned-from-chinas-4th-plenum/.

Transparency International's Corruption Perception Index for 2016 ranked China 79th out of 176 countries and territories, up from 72nd in 2007.[86]

Demographic Challenges

Many economists contend that China's demographic policies, particularly its one-child policy (first implemented in 1979), are beginning to have a significant impact on the Chinese economy. For example, according to a McKinsey Global Institute study, China's fertility rate fell from about 5.8 births per woman in 1964 to 1.6 in 2012.[87] This is now affecting the size of the Chinese workforce.

The existence of a large and underemployed labor force was a significant factor in China's rapid economic growth when economic reforms were first introduced. Such a large labor force meant that firms in China had access to a nearly endless supply of low-cost labor, which helped enable many firms to become more profitable, which in turn led them to boost investment and production. Some economists contend that China is beginning to lose this labor advantage. According to the Chinese government, the size of its working age population (ages 16 to 59) peaked at 925 million in 2011, but then fell for seven consecutive years to 897 million in 2018. The Chinese government projects that its working age population will drop to 830 million by 2030 and to 700 million by 2050. If these projections prove accurate, the Chines working age population could drop by 225 million individuals (2011-2050).[88]

The one-child policy has also resulted in a rapidly aging society in China.[89] According to the Brookings Institute, China already has 180 million people aged over 60, and this could reach 240 million by 2020 and

---

[86] Transparency International, Corruption Perceptions Index 2016, January 2017, available at https://www.transparency.org/news/feature/corruption_perceptions_index_2016.

[87] McKinsey Global Institute, *Can Long-Term Global Growth be Saved?,* January 2015, available at http://www.mckinsey.com/~/media/McKinsey/dotcom/Insights/Growth/Can%20long-term%20global%20growth%20be%20saved/MGI_Global_growth_Full_report_February_2015pdf.ashx.

[88] Caixing, *Chart of the Day: China's Shrinking Workforce*, January 29, 2019, available athttps://www.caixinglobal.com/2019-01-29/chart-of-the-day-chinas-shrinking-workforce-101375782.html.

[89] Some analysts contend that because of it demographics, China will grow old before it grows rich.

360 million by 2030. The population share of people aged over 60 could reach 20% by 2020, and 27% by 2030.[90] With a declining working population and a rising elderly population, the Chinese government will face challenges trying to boost worker productivity (such as enhancing innovation and high-end technology development) and expanding spending on health care and elderly services. China's Hukou (household registration) system also poses challenges to the government.

China's Hukou System[91]

First introduced in 1951, the Chinese Hukou (household registration) system is a categorization of its citizens based on both their place of residence and eligibility for certain socioeconomic benefits. Hukou is issued through a registration process administered by local authorities and solidified into inheritable social identities.[92] The classification of the system is composed by two related parts: socioeconomic eligibility (agriculture/nonagriculture); and residential location (living in urban/rural areas). The Chinese government imposed the system with the purpose of regulating population distribution, especially in regard to cities. Since economic reforms were begun in 1979, hundreds of millions of people have been allowed to leave their home towns to work in urban areas, such as Shanghai. The number of rural laborers working in China's cities was 274 million in 2014, over one-third (36%) of the total workforce.[93] Although such workers are allowed to reside in the cities where they work, they are generally denied access to social entitlements, such as pensions, medical insurance, and basic education for children. This forces such workers to save a very high level of their income to pay for these services. Due to China's desire to increase the urbanization of its population, combat demographic disparities, and boost domestic consumption, the Chinese government is currently considering implementing new reforms to the Hukou system.

Economic Goals of the 19th Party Congress of the Communist Party

President Xi's report to the 19th Party Congress in November 2017 stated that socialism with Chinese characteristics had entered a new era. He

---

90 Brookings Institute, *Racing Towards the Precipice*, by: Feng Wang, June 2012, available at http://www.brookings.edu/research/articles/2012/06/china-demographics-wang.

91 Prepared by Candy Meza, Research Associate, Foreign Affairs, Defense, and Trade Division.

92 Congressional-Executive Commission on China, *Special Topic Paper: China's Household Registration System: Sustained Reforms Needed to Protect China's Rural Migrant.*

93 Annual survey of migrant workers conducted by the National Bureau of Statistics, 2014. http://www.stats.gov.cn/tjsj/ ndsj/2014/indexeh.htm.

stated that China would work to become a "moderately prosperous society in all respects" by 2050. Major goals include boosting living standards for poor and rural people, addressing income disparities (e.g., rich-poor and urban-rural), making private consumption the driver of the economy, boosting services, reducing pollution, promoting innovation and economic modernization, and improving overall living standards.[94] For example, the report states the following:

> We will work faster to build China into a manufacturer of quality and develop advanced manufacturing, promote further integration of the internet, big data, and artificial intelligence with the real economy, and foster new growth areas and drivers of growth in medium-high end consumption, innovation-driven development, the green and low-carbon economy, the sharing economy, modern supply chains, and human capital services. We will support traditional industries in upgrading themselves and accelerate development of modern service industries to elevate them to international standards. We will move Chinese industries up to the medium-high end of the global value chain, and foster a number of world-class advanced manufacturing clusters.

The report indicated that China would continue to pursue trade and investment reforms, noting the following:

> We will adopt policies to promote high-standard liberalization and facilitation of trade and investment; we will implement the system of pre-establishment national treatment plus a negative list across the board, significantly ease market access, further open the service sector, and protect the legitimate rights and interests of foreign investors. All businesses registered in China will be treated equally.

However, the report emphasized the continued importance of the state sector and the government's continued role in various economic sectors:

---

[94] China's goals are to achieve average annual GDP growth of 4.8% from 2020 to 2035 and 3.4% from 2030 to 2050. It seeks to achieve per capita GDP of $20,000 by 2025 (making China a high income country), $45,000 by 2035 (35% of U.S. levels), and $120,000 by 2050 (half of U.S. levels).

> We will improve the systems for managing different types of state assets, and reform the system of authorized operation of state capital. In the state-owned sector, we will step up improved distribution, structural adjustment, and strategic reorganization. We will work to see that state assets maintain and increase their value; we will support state capital in becoming stronger, doing better, and growing bigger, and take effective measures to prevent the loss of state assets. We will further reform of state-owned enterprises, develop mixed-ownership economic entities, and turn Chinese enterprises into world-class, globally competitive firms.[95]

China's Belt and Road Initiative

China's Belt and Road initiative (BRI), also called "One Belt, One Road" (OBOR), was launched in 2013 to boost economic integration and connectivity (such as infrastructure, trade, and investment) with its neighbors and various trading partners in Asia, Africa, Europe, and beyond.[96] At the APEC summit in November 2017, President Xi said the following:

> The Belt and Road Initiative calls for joint contribution and it has a clear focus, which is to promote infrastructure construction and connectivity, strengthen coordination on economic policies, enhance complementarity of development strategies and boost interconnected development to achieve common prosperity. This initiative is from China, but it belongs to the world. It is rooted in history, but it is oriented toward the future. It focuses on the Asian, European and African continents, but it is open to all partners. I am confident that the launch of the Belt and Road Initiative will create a broader and more dynamic platform for Asia-Pacific cooperation.[97]

---

[95] Xinhua, "Full text of Xi Jinping's report at 19th CPC National Congress," November 3, 2017, available at http://news.xinhuanet.com/english/special/2017-11/03/c_136725942.htm.

[96] It comprises the Silk Road Economic Belt and the 21st Century Maritime Silk Road.

[97] Xinhuanet, "Full text of Chinese President Xi's address at APEC CEO Summit," November 11, 2017, available at http://www.xinhuanet.com/english/2017-11/11/c_136743492.htm.

Many U.S. analysts view the BRI differently than how Chinese leaders describe it. For example, Nadège Rolland, senior fellow with the National Bureau of Asian Research states the following:

> The Belt and Road Initiative (BRI) is generally understood as China's plan to finance and build infrastructure projects across Eurasia. Infrastructure development is in fact only one of BRI's five components which include strengthened regional political cooperation, unimpeded trade, financial integration and people-to-people exchanges. Taken together, BRI's different components serve Beijing's vision for regional integration under its helm. It is a top-level design for which the central government has mobilized the country's political, diplomatic, intellectual, economic and financial resources. It is mainly conceived as a response to the most pressing internal and external economic and strategic challenges faced by China, and as an instrument at the service of the PRC's vision for itself as the uncontested leading power in the region in the coming decades. As such, it is a grand strategy.[98]

Many aspects of the BRI initiative remain unclear, including which (and how many) countries will participate, how much China will spend to finance the initiative, and what projects will fall under the BRI. For example, the government's China Belt and Road Portal currently lists profiles of 70 countries on its website.[99] However, China's official media in December 2017 stated that 86 countries and international organizations had signed 100 cooperation agreements with China under the BRI.[100] Nadège Rolland said that China pledged it would spend $1 trillion to $1.3 trillion, *The Economist* reports that China put the figure at $4 trillion,[101]

---

[98] Testimony of Nadège Rolland, Senior Fellow, The National Bureau of Asian Research, before the U.S.-China Economic and Security Review Commission Hearing on: "China's Belt and Road Initiative: Five Years Later, January 25, 2018, available at https://www.uscc.gov/sites/default/files/Rolland_USCC%20Testimony_16Jan2018.pdf.

[99] See https://eng.yidaiyilu.gov.cn/info/iList.jsp?cat_id=10076&cur_page=1.

[100] Xinhuanet, "China signs cooperation agreements with 86 entities under Belt and Road," available at http://www.xinhuanet.com/english/2017-12/23/c_136846221.htm.

[101] The Economist, "Our bulldozers, our rules, China's foreign policy could reshape a good part of the world economy," July 2, 2016, available at https://www.economist.com/news/china/21701505-chinas-foreign-policy-couldreshape-good-part-world-economy-our-bulldozers-our-rules.

and the World Economic Forum estimates that China could ultimately spend $8 trillion on BRI.[102]

The initiative could provide a big boost to China's economy and soft power image. China hopes to gain a better return on its foreign exchange reserves, create new overseas business opportunities for Chinese firms, create new markets for industries currently experiencing overcapacity, and stimulate economic development in poorer regions of China.[103] However, the initiative could pose financial risks if borrowers do not repay loans or if recipient countries do not view Belt and Road as benefiting them. U.S. Secretary of State Rex Tillerson criticized certain aspects of Belt and Road initiative in remarks made in October 2017:

> We have watched the activities and actions of others in the region, in particular China, and the financing mechanisms it brings to many of these countries which result in saddling them with enormous levels of debt. They don't often create the jobs, which infrastructure projects should be tremendous job creators in these economies, but too often, foreign workers are brought in to execute these infrastructure projects. Financing is structured in a way that makes it very difficult for them to obtain future financing, and oftentimes has very subtle triggers in the financing that results in financing default and the conversion of debt to equity.[104]

China has undertaken other major financial initiatives as well. In July 2014, China, along with Brazil, Russia, India, and South Africa, announced the creation of a $100 billion "New Development Bank," which is headquartered in Shanghai, China. The new bank aims to fund infrastructure projects in developing countries. In October 2014, China launched the creation of a new $100 billion Asian Infrastructure Development Bank (AIIB), aimed at funding infrastructure projects in

---

102 World Economic Forum, "China's $900 billion New Silk Road. What you need to know," June 26, 2017, available at https://www.weforum.org/agenda/2017/06/china-new-silk-road-explainer/.

103 On October 24, 2017, the 19th Chinese Communist Party Congress passed a resolution to include the Belt and Road Initiative into the Chinese Constitution.

104 U.S. Department of State, Remarks on "Defining Our Relationship with India for the Next Century," October 18, 2017, available at https://www.state.gov/secretary/remarks/2017/10/274913.htm.

Asia.[105] Fifty-seven nations joined as founding members. The AIIB, headquartered in Beijing, announced it was open for business in January 2016. To date, the United States has chosen not to join the AIIB.

Made in China 2025

The "Made in China 2025" initiative, announced in 2015, is one of several recently announced ambitious projects aimed at increasing the competitiveness of Chinese industries, fostering Chinese brands, boosting innovation, and reducing China's reliance on foreign technology by making China a major or dominant global manufacturer of various technologies.[106] According to Chinese media, the initiative intends to "transform China from a manufacturing giant into a world manufacturing power" by 2049.[107] For example, the plan states a goal of achieving 40% of domestically manufactured basic components and basic materials by 2020 and 70% by 2025. An updated version of the plan released in January 2018 said China aimed to become the world's leading manufacturer of telecommunication, railway, and electrical power equipment by 2025, and that China's robotics, high-end automation, and new energy vehicles industries would globally rank second or third by 2025.[108] The methods the Chinese government plans to use to achieve its goals have raised concerns among U.S. firms and policymakers because they appear to involve large subsidies, protection of domestic industries, directed policies to purchase technology and IPR from abroad, increased pressure on foreign firms to

---

[105] See CRS In Focus IF10154, *Asian Infrastructure Investment Bank*, by Martin A. Weiss.

[106] The 2015 "Made in China 2025" document identified these 10 for support These ten key sectors are (1) next-generation information technology, (2) high-end numerical control machinery and robotics, (3) aerospace and aviation equipment, (4) maritime engineering equipment and high-tech maritime vessel manufacturing, (5) advanced rail equipment, (6) energy-saving and new energy vehicles, (7) electrical equipment, (8) agricultural machinery and equipment, (9) new materials, and (10) biopharmaceuticals and high-performance medical devices.

[107] Xinhuanet, "*Made in China 2025" Plan Unveiled*, May 19, 2015, at http://www.xinhuanet.com/english/2015-05/19/ c_134251770.htm

[108] China Daily, "Made in China 2025 roadmap updated," January 27, 2018, at http://www.chinadaily.com.cn/a/ 201801/27/WS5a6bb8b9a3106e7dcc137168.html.

transfer technology in order to do business in China, and what appears to be a goal of deliberately reducing foreign participation in China's markets.

In an interview on November 3, 2017, U.S. Trade Representative Robert Lighthizer stated that China's Made in China 2025 initiative was "a very, very serious challenge, not just to us, but to Europe, Japan and the global trading system."[109] The USTR's 2017 annual report on China's WTO compliance focused heavily on the initiative, stating that Made in China 2025 differed from industry support by other WTO members in the level of ambition and scale of resources dedicated to obtaining its goals, and the USTR report warned that "even if the Chinese government fails to achieve the industrial policy goals set forth in Made in China 2025, it is still likely to create or exacerbate market distortions and create severe excess capacity in many of the targeted industries."[110]

Challenges to U.S. Policy of China's Economic Rise

China's rapid economic growth and emergence as a major economic power have given China's leadership increased confidence in its economic model. Many believe the key challenges for the United States are to convince China that (1) it has a stake in maintaining the international trading system, which is largely responsible for its economic rise, and should take a more active leadership role in maintaining that system; and (2) further economic and trade reforms are the surest way for China to grow and modernize its economy. Lowering trade and investment barriers would boost competition in China, lower costs for consumers, increase economic efficiency, and spur innovation. However, many U.S. stakeholders are concerned that China's efforts to boost the development of indigenous innovation and technology could result in greater intervention

---

[109] The interview is available at https://www.youtube.com/watch?v=L03Np5ZLvM8.

[110] USTR, 2017 Report to Congress on China's WTO Compliance, January 2018, p. 10, available at https://ustr.gov/sites/default/files/files/Press/Reports/China%202017%20WTO%20Report.pdf.

by the state (such as subsidies, trade and investment barriers, and discriminatory policies), which could negatively affect U.S. IP-intensive firms.

Opinions differ as to the most effective way to deal with China on major economic issues. Some support a policy of engagement with China using various forums. Others support a somewhat mixed policy of using engagement when possible, coupled with a more aggressive use of the WTO dispute settlement procedures to address China's unfair trade policies.[111] Others, who see China as a growing threat to the U.S. economy and the global trading system, advocate a policy of trying to contain China's economic power and using punitive measures, such as increased tariffs under Section 301, to either counter the negative impact of China's industrial policies on U.S. firms or push China to modify distortive and discriminatory policies (such as the Made in China 2025 initiative). Responding to China's BRI is viewed by some as a major challenge to U.S. global economic interests. While China's financial support of infrastructure projects in numerous countries could produce positive economic results, U.S. policymakers have expressed concerns that China will use BRI to mainly benefit its own firms, that the process of implementation of projects will not be transparent, that BRI participation could saddle countries with large debts, and that China will use the BRI to spread its economic system to other countries.

[111] At the time of this writing, the Trump Administration has not brought any WTO dispute settlement cases against China to date.

In: Key Congressional Reports for June 2019 ISBN: 978-1-53616-666-8
Editor: Tara T. Robertson

Chapter 2

INDIA'S 2019 NATIONAL ELECTION AND IMPLICATIONS FOR U.S. INTERESTS*

K. Alan Kronstadt

ABSTRACT

India, a federal republic and the world's most populous democracy, held elections to seat a new lower house of parliament in April and May of 2019. Estimates suggest that more than two-thirds of the country's nearly 900 million eligible voters participated. The 545-seat Lok Sabha (People's House) is seated every five years, and the results saw a return to power of the Bharatiya Janata Party (BJP) led by Prime Minister Narendra Modi, who was chief minister of the west Indian state of Gujarat from 2001 to 2014. Modi's party won decisively—it now holds 56% of Lok Sabha seats and Modi became the first Indian leader to win consecutive majorities since Indira Gandhi in 1971.

The United States and India have been pursuing an expansive strategic partnership since 2005. The Trump Administration and many in the U.S. Congress welcomed Modi's return to power for another five-year term. Successive U.S. Presidents have deemed India's growing power and

* This is an edited, reformatted and augmented version of Congressional Research Service Publication No. R45807, dated June 28, 2019.

influence a boon to U.S. interests in Asia and globally, not least in the context of balancing against China's increasing assertiveness. India is often called a preeminent actor in the Trump Administration's strategy for a "free and open Indo-Pacific." Yet there are potential stumbling blocks to continued development of the partnership. In 2019, differences over trade have become more prominent, and India's long-standing (and mostly commercial) ties to Russia and Iran may run afoul of U.S. sanctions laws. Additionally, India maintains a wariness of U.S. engagement with Pakistan and intentions in Afghanistan, with Islamabad presently facilitating a U.S.-Taliban dialogue and India counseling against a precipitous U.S. withdrawal from Afghanistan.

Prime Minister Modi's return to power promises broad continuity, even with some notable changes to the federal cabinet. By many accounts, Modi's record as an economic reformer and liberalizer is mixed, and his reputation as a nationalist "watchman" has not always translated into effective foreign policy, according to some analysts. It is unclear if Modi will use his renewed domestic political mandate to pursue more assertiveness internationally, possibly in ways that challenge U.S. preferences. Still, most analysts contend that Modi and the BJP have been and will continue to be more open to aligning with U.S. regional strategy and more energetic in pursuing U.S.-favored economic reforms than would have been any alternative Indian leadership.

The BJP is a Hindu nationalist party, born in 1980 of a larger social movement, and Narendra Modi is a self-avowed Hindu nationalist (India is roughly 79% Hindu and 14% Muslim). The 2019 Modi-BJP campaign was widely criticized for divisiveness, and nationalist fervor following a February India-Pakistan crisis may have benefitted the BJP at the polls. India's minority communities and the country's civil society are widely reported to be under increasing threats emanating from Hindu majoritarian policies and sentiment. These threats can take violent and repressive forms, at times with the involvement of Indian officials or political figures, as reported by the U.S. State Department and independent human rights watchdogs, and as criticized by some Members of Congress.

This chapter reviews the recent Indian election process and results, the country's national political stage, and possible implications for U.S. interests in the areas of bilateral economic and trade relations, defense and security ties, India's other foreign relations, and human rights concerns.

Overview

India is a federal parliamentary republic and the world's most populous democracy, with more than 1.3 billion citizens. In the spring of 2019 the country held elections to seat its 17th Lok Sabha (House of the People), the 545-seat lower chamber of parliament and locus of Indian political power. Results were announced on May 23, with the incumbent Bharatiya Janata Party (BJP, or Indian Peoples Party) winning a sweeping and repeat victory under Prime Minister Narendra Modi. Having in 2014 become the first party to attain a parliamentary majority (52%) in 30 years, the BJP was able to expand that majority to 56% in 2019, becoming the first party to win consecutive majorities since 1971. The dynastic Indian National Congress (hereinafter, Congress Party)—which had dominated the country's politics from 1947 to 1977 and led a national coalition government from 2004 to 2014—again failed to win the 10% minimum of seats required to officially lead the Lok Sabha opposition. Powerful regional and caste-based parties likewise posed no meaningful obstacles to the latest "BJP wave."

The Administration of President Donald Trump is seeking to expand upon a "strategic partnership" with India formally launched in 2005. Progress is ongoing, most notably in the area of defense and security cooperation.[1] The U.S. Congress has been supportive of efforts to expand and deepen the bilateral partnership, formally designating India as a "Major Defense Partner" of the United States in 2016. Areas of engagement are broad and include an array of economic and security initiatives. Recent bilateral frictions have arisen over trade practices, religious freedom, and India's relations with Russia and Iran.

A State Department release congratulated Prime Minister Modi and his ruling National Democratic Alliance (NDA) coalition for their "decisive victory," and it applauded the Indian people for turning out in historic numbers and the Indian government for "exceptional execution of this massive undertaking." It went on to declare that

---

[1] See, for example, the September 6, 2018, Joint Statement at https://go.usa.gov/xmMJj.

> The United States and India enjoy a strong strategic partnership that stands on a foundation of shared values, extensive people-to-people ties, and a commitment to a secure and prosperous Indo-Pacific region. We look forward to working with the newly elected government on a range of important issues.... We are confident that the strong and upward trajectory of our partnership will continue.[2]

The Chairman of the House Foreign Affairs Committee also issued a laudatory statement and, in June, several Members of Congress penned an open letter to President Trump to "highlight the continuing strategic importance of the U.S.-India relationship."[3] Secretary of State Michael Pompeo met with Modi and other top Indian officials in New Delhi in June to express enthusiasm about America's "natural strategic partner." President Trump has yet to visit India, and high-level engagement occurs under the rubric of a "2+2 Ministerial Dialogue" inaugurated in September 2018.[4] Nevertheless, many independent observers express concern about emergent frictions and indications that the partnership is becoming dissonant. A recent substantive overview of the relationship concluded that it "would benefit from some realism about its limitations."[5]

Officials in governments across the Indo-Pacific region expect the election results to bring general continuity in Indian policies. Given the instability and uncertainty that can accompany coalition governance, many likely regarded Prime Minister Modi's convincing reelection with some relief.[6] Modi's internationalist orientation can be seen in his energetic pursuit of diplomacy with numerous major powers and regional governments. This includes though continued development of the U.S.-

---

[2] See the May 24, 2019, State Department release at https://www.state.gov/indian-elections.

[3] See the Chairman's statement at https://go.usa.gov/xy2QJ, and the June 27, 2019, letter at https://go.usa.gov/xy2Qv.

[4] State Department's June 25, 2019, release at https://go.usa.gov/xy2E4.

[5] Sameer Lalwani and Heather Byrne, "The Elephant in the Room: Auditing the Past and Future of the U.S.-India Partnership," War on the Rocks (online), June 26, 2019. See also Tanvi Madan, "As Pompeo Heads to Delhi, the US-India Relationship is at a Critical Juncture," Brookings Institution, June 21, 2019; "What Is Behind the Tensions in US- India Relations?," *Financial Times* (London), June 24, 2019.

[6] Ian Hall, "After Modi's Election Win, What's Next for India in the Indo-Pacific?," Australian Strategic Policy Institute (Canberra), May 24, 2019.

India strategic partnership, which is widely seen to be rooted in an array of mutually-held values and increasingly convergent visions for global order, lately conceived by Washington as a "shared vision for a free, open, and rules-based Indo-Pacific region."[7]

On the economic front, Modi's reputation as a reformer and liberalizer has met with mixed reviews. Five years in office realized no major land or labor reforms or meaningful efforts to address bad bank debt that were widely anticipated. The BJP leader's assumed electoral vulnerabilities—relatively lackluster economic expansion (averaging below 7% annually), joblessness, rising prices, and a widely panned 2016 "demonetization" initiative—did not end up degrading his impressive political support. Most market-oriented analysts are hopeful, but not entirely confident, that a "Modi 2.0" government will redouble efforts to implement the kinds of economic reforms sought by the U.S. and other governments and business interests.[8]

ELECTION AND OUTCOME

India possesses a robust and competitive multiparty democratic system.[9] Yet its politics also are described as "beset by corruption" by Freedom House, a think tank that ranks world countries on levels of political and civil liberties. In 2017, India had fallen 10 places on the Economic Intelligence Unit Democracy Index to 42nd in the world, due in part to "a rise of vigilantism and violence against minority communities, particularly Muslims, as well as other dissenting voices." For 2018, India was ranked 41st worldwide, but its overall score was unchanged, and it

---

[7] See the State Department's June 25, 2019, release at https://go.usa.gov/xy4Mv.

[8] An assessment of the Modi government's performance on the 30 "significant economic reforms" most sought by trade bodies, economists, and rating agencies found its track record to be "solid" since 2014, with 9 of the reforms fully enacted and another 15 partially complete ("U.S.-India Insight: Scoring the Modi Government's Reform Program," Center for Strategic and International Studies, April 22, 2019).

[9] See CRS In Focus IF10298, *India's Domestic Political Setting*, by K. Alan Kronstadt.

maintained its designation as a "flawed democracy" (as did the United States).[10]

The scale of India's national election presents daunting logistical challenges, and voting was completed in seven phases. As with each iteration, the event was history's largest democratic exercise: About 880 million people—one-ninth of the world's population—were eligible to cast ballots in the country's 29 states and 7 Union Territories, including some 84 million first-time voters. More than 600 million Indians participated, for a turnout rate of 67.4%. More than 8,400 candidates and a record 669 parties vied for the Lok Sabha's 543 elected seats; 36 parties won at least one seat, as did 4 independent candidates.[11]

In addition to being history's largest democratic undertaking, India's 2019 national election was also history's most expensive: participating parties and candidates spent an estimated $8.7 billion on the campaign, more than double the 2014 spending. The BJP reportedly received nearly three-quarters of all political donations. Such massive outlays raise concerns among many about fairness: there is little transparency in India's campaign finance system, the Modi government had lifted caps on corporate donations, and the Election Commission's oversight is criticized for ineffectiveness.[12]

In the lead-up to the 2019 voting, analysts debated whether the BJP's 2014 sweep was an anomaly in a decades-long era of coalition politics in New Delhi, or marked the beginning of a new period of single-party domination as was seen under the Congress Party prior to 1989. The BJP's unexpected and even stronger showing in 2019 suggests that India's

---

[10] See https://freedomhouse.org/report/freedom-world/2019/india and http://www.eiu.com/topic/democracy-index.

[11] Ashoka University Trivedi Center for Political Data (Sonepat) at http://lokdhaba.ashoka.edu.in.

[12] "Lok Sabha Elections 2019: Analysis of Criminal Background, Financial, Education, Gender, and Other Details of Winners," Association for Democratic Reforms (New Delhi), May 25, 2019; "India's Elections: Rules on Campaign Financing and Spending," Reuters, April 30, 2019.

national political stage has undergone qualitative change back toward a hegemonic national party.[13]

Table 1. Leading parties in India's 2019 National Election (percentages are rounded)

Party	Seats Won	% of Seats Won	% of Popular Vote
Bharatiya Janata Party (BJP)	303	56%	37%
Indian National Congress	52	10%	19%
Dravida Munnetra Kazhagam (Tamil Nadu)	23	4%	2%
All India Trinamool Congress (West Bengal)	22	4%	4%
YSR Congress (Andhra Pradesh)	22	4%	3%
Shiv Sena (Maharashtra)	18	3%	2%
Janata Dal (United) (Odisha)	16	3%	2%
All others	87	16%	31%

Source: Election Commission of India (http://results.eci.gov.in/pc/en/partywise/ index. htm).

The BJP won by securing 303 seats with more than 37% of the aggregate vote, significantly exceeding the 272 seats required for majority status, and increasing the party's seat total from 282 in the previous Lok Sabha. Their closest competitors, the Congress Party, took 52 seats with 19.5% of the national vote, a net gain of 8 seats (see *Table 1*). According to Election Commission of India data, the BJP won a majority of votes cast in 10 states, as well as in the Delhi National Capital Territory. It also took 49.6% of the votes cast in Uttar Pradesh (UP), India's most populous state.[14] No other party was able to garner a majority of votes in any single state. While the BJP's core support in India's western and north-central "Hindi belt" states remained strong, and the party succeeded in expanding its appeal in the country's east, it was shut out completely in seven states,

---

[13] Milan Vaishnav, "Is the BJP India's New Hegemon?," Carnegie Endowment for International Peace, October 8, 2018; "How Narendra Modi Has Reinvented Indian Politics," BBC News, May 23, 2019.

[14] With more than 200 million citizens—about one-fifth of them Muslim—Uttar Pradesh is India's leading political bellwether state (see Milan Vaishnav, "As Uttar Pradesh Goes, So Goes India," Carnegie Endowment for International Peace, February 5, 2019).

including the major southern states of Andhra Pradesh and Tamil Nadu, where regional parties prevailed. In a sign of the Congress Party's historic collapse nationally, the party failed to win a single seat in 14 states, and nearly half of the seats the party did win (23) came from India's two southernmost states, Kerala and Tamil Nadu (see *Figure 1*, "Map of Indian States").

Selected Characteristics of 17th Lok Sabha Members

- 88% are *crorepati* (Indian "millionaires")
- 73% have a college degree
- 55% are first-time members
- 43% face declared criminal cases (see footnote 16)
- 30% are "dynasts" (from political families)
- 29% are upper-caste Hindus
- 15% are females (up from 12% previously)
- 12% are under age 40 (up from 8%)
- 4.6% are Muslims (versus 14% of the population)

Sources: Parliamentary Research Service (New Delhi); Ashoka University Trivedi Center for Political Data (Sonepat); Association for Democratic Norms (New Delhi)
Note: A "crorepati" is a person from a household with assets of at least 10 million rupees (approximately $143,000)

The newly seated Lok Sabha is, on average, younger, wealthier, and better educated than its predecessor, and the number of female members has increased to 15% (see *Text Box*). Upper-caste Hindus and political families continue to enjoy disproportionately high representation, and India's large and relatively marginalized Muslim minority community of about 190 million (about 14% of the total) suffers from a declining voice in Parliament: Muslim representation in the Lok Sabha peaked at 10% in 1980 and lingered at about 6% until dropping to 4% in 2014. A net gain of three Muslim members in 2019 only slightly raised that percentage.[15]

[15] "Fact Sheet: Muslim Representation in Parliament," *India Today* (Delhi), March 10, 2014.

India's parliamentarians are also notable for the numbers who face formal criminal allegations.[16]

Notable Changes in Union Ministers

The "Modi 2.0" cabinet includes some significant changes to the leadership of key Indian ministries. For the first time ever, a career diplomat, Subramanyam Jaishankar, is now External Affairs Minister, replacing BJP stalwart Sushma Swaraj. Jaishankar, until recently Foreign Secretary, is a former ambassador to both the United States and China, and is widely known in Washington as proponent and facilitator of closer U.S.-India ties. Another key new figure in the Modi cabinet is Home Minister Amit Shah, who succeeds senior BJP official Rajnath Singh, himself now holding the Defense Ministry portfolio. Also under new leadership is the Finance Ministry, where former Defense Minister Nirmala Sitharaman has become India's first female finance minister. Additionally, National Security Advisor Ajit Doval, a former intelligence chief in office since 2014 and considered a hardliner on Pakistan, was elevated to cabinet rank in 2019.

Other Parties and Figures

The Indian National Congress

India's Congress Party had hoped in 2019 to reverse its general decline after 2014. It had led a United Progressive Alliance coalition government

---

[16] Members of the 17th Lok Sabha are significantly more likely to have criminal histories than those in the 16th: more than two in five (43%) of the winning candidates in 2019 reportedly face criminal charges, including nearly 29% facing serious charges such as murder or rape. There also is evidence that, at the national and state level, BJP politicians are three times as likely as Congress figures to be linked to crimes involving religious violence ("Lok Sabha Elections 2019: Analysis of Criminal Background, Financial, Education, Gender, and Other Details of Winners," Association for Democratic Reforms (New Delhi), May 25, 2019; "Let's Talk About Hate: In Indian Politics, Candidates Who Stoke Communal Hatred Thrive," *Hindustan Times* (Delhi), July 27, 2017).

under Prime Minister Manmohan Singh from 2004 to 2014. Party president Rahul Gandhi—son of former Prime Minister Rajiv Gandhi and grandson of Indira Gandhi—had entered politics with apparent reluctance, winning the "family seat" in Amethi, UP, in 2004. Ten years later, Rahul oversaw Congress's worst electoral performance in history when the party took only 44 Lok Sabha seats, down from 206 previously.

Gandhi's lackluster reputation took a new shine after the party's unexpectedly strong showing in the 2017 Gujarat state elections, and December 2018 Congress victories in three north-central states were a huge morale boost for the party, heartening potential Congress allies. During the 2019 campaign, Gandhi impressed many observers with what were described as creative policy proposals and a newly assured public speaking style.[17] Still, going into the 2019 campaign, Congress had no significant presence in any of India's six most populous states and had shown an inability to recover in states where it had faltered. The party continued to struggle to both consolidate old alliances and to establish new ones.[18] As put by one official from a BJP-allied regional party, the opposition had "neither a program, nor a leader, nor a narrative." Gandhi suffered an embarrassing upset loss in his family's Amethi district, and the party's 2019 defeat has led to new leadership crises.[19]

Regional and Caste-based Parties

The BJP and Congress are India's only truly national parties—they together won roughly half of all votes cast in 2009 and 2014, and their combined share rose to 57% in 2019 (attributable to a 6- point boost for the BJP). The influence of regional and caste-based parties—although blunted by the BJP's outright majority victories—remains a crucial variable in Indian politics. Such parties now hold about one-third of Lok Sabha seats, but many of the most influential met with significant reversals in 2019.

---

[17] "Should Narendra Modi Be Worried?," BBC News, December 12, 2018; Sushil Aaron, "The Remarkable Comeback of Rahul Gandhi" (op-ed), *New York Times*, April 22, 2019.

[18] Milan Vaishnav, "Is the BJP India's New Hegemon?," Carnegie Endowment for International Peace, October 8, 2018; Ajoy Ashirwad Mahaprashasta, "As the Alliance Game Continues, Congress's Missteps Could Cost It Dearly" (op-ed), *The Wire* (Delhi), March 15, 2019.

[19] Quote in "Indian Election Results: Modi Wins a Landslide Victory," *Washington Post*, May 23, 2019.

By early 2019, Narendra Modi had become the primary, if not sole target of his many electoral opponents, but no single challenger emerged. Still, the opposition's zeal to dislodge the incumbents had them ready to make unusual alliances, especially in Uttar Pradesh, where the effort failed conclusively.[20] Other powerful regional parties experienced setbacks, most notably West Bengal's Trinamool Congress, led by Chief Minister Mamata Banerjee, which barely survived a surprise BJP surge in the key eastern state, winning 22 of 42 Lok Sabha seats to the BJP's 18 (up from 2 in 2014). Only in Odisha was a major regional party—in this case the Biju Janata Dal of popular Chief Minister Naveen Patnaik—able to withstand the BJP onslaught.

IMPLICATIONS FOR U.S. INTERESTS

Because the BJP campaign was run largely on Narendra Modi's personal popularity rather than an explicit policy platform, it is unclear how Modi will use his mandate going forward. Since 2014, the Modi government arguably has realized some foreign policy successes compatible with U.S. interests: sustaining the partnership with the United States, solidifying the partnership with Japan, strengthening ties with Israel while making new outreach to key Persian Gulf states such as Saudi Arabia and the UAE, and articulating a vision for the Indo-Pacific region that tracks well with that of the United States.[21] Modi has successfully projected India as the world's next big economic opportunity after China, but critics argue that he has mostly squandered an opportunity to move

---

[20] In January 2019, UP's two major caste-based parties and often bitter rivals, the Bahujan Samaj Party and Samajwadi Party, announced they would cooperate in upcoming national elections with the aim of defeating the BJP in their state. The alliance, which did not include Congress, was seen by some as an act of honorable statesmanship, but, ultimately, even in tandem the two parties were unable to reverse the BJP's new dominance in the state, fueling doubt about the efficacy of "caste politics" in UP ("Get Modi: 'Blood Rivals' Bury the Hatchet to Beat India's Leader," *Wall Street Journal*, March 5, 2019; "Is It the End of Caste Politics in Uttar Pradesh?," *Economic Times* (Mumbai), May 23, 2019).

[21] Ashley Tellis, "Troubles Aplenty: Foreign Policy Challenges for the Next Indian Government," Carnegie Endowment for International Peace, May 20, 2019.

India into great power status, with a lack of strategic vision harming India's position vis-à-vis major powers and smaller neighbors, alike.[22] Given India's myriad domestic problems, and still-limited capacity to project power, some American observers are skeptical about its near-term potential to play the role sought for it by the U.S. Congress and successive Administrations.[23]

The Modi/BJP victory has empowered the Indian leader domestically and this may provide Modi and India new opportunities on the global stage. Given Modi's reputation for favoring a "muscular" foreign policy, he may now be more willing to resist Chinese assertiveness and move closer to the United States.[24] Yet troubles with the United States also could loom: Many Indian strategic thinkers say their country's national interests are well served by engaging not just with the United States but also with Russia and Iran, which could limit to New Delhi's willingness to abide what some Indian observers describe as "America's short-term impulses."[25] While New Delhi generally welcomes the U.S. "free and open Indo-Pacific" (FOIP) strategy, Indian leaders continue to demur from confronting China. The United States and India also seek to cooperate on energy, climate change, and space issues, and have sometimes clashing views on immigration.

---

[22] Shayam Saran, "In Year Four, Modi's Foreign Policy Needs Some Course Correction," (op-ed), *The Wire* (online, New Delhi), May 29, 2017; Shivshankar Menon, "The Reality of Narendra Modi's Foreign Policy Failures Laid Bare" (book review), *The Wire* (online, Delhi), November 3, 2018.

[23] Michael Swaine, "Creating an Unstable Asia: The U.S. 'Free and Open Indo-Pacific' Strategy," Carnegie Endowment for International Peace, March 2, 2018; Daniel Byman, "How India's Rise Can Complement U.S. Strategy," Lawfare (online), October 7, 2018; Sameer Lalwani and Heather Byrne, "The Elephant in the Room: Auditing the Past and Future of the U.S.-India Partnership," War on the Rocks (online), June 26, 2019.

[24] Brahma Chellaney, "Election Triumph Will Boost Modi's International Clout," *Nikkei Asian Review* (Tokyo), June 1, 2019; Richard Fontaine, "Modi's Victory Is America's Opportunity," *National Review*, May 29, 2019.

[25] Samir Saran, "'In India We Trust' Would Be Good US Policy," Observer Research Foundation (New Delhi), June 26, 2019. See also Manoj Joshi, "Dealing With the Big Bully" (op-ed), *Tribune India* (Chandigarh), June 25, 2019.

Bilateral Trade and Economic Relations[26]

Trade and economic ties, an important and growing part of bilateral relations, have faced recent challenges.[27] Prime Minister Modi's strong electoral mandate suggests India's mixed economic performance did not hurt his standing with the public, and the results may embolden the BJP to press ahead with its reform agenda with greater vigor (see *Text Box*, below).[28] The Trump Administration takes issue with the U.S. trade deficit with India and "unfair" trade practices that restrict U.S. exports to and investment in India. Some U.S. policymakers and businesses have been disappointed that, during Modi's first term, India did not move forward with market-opening reforms as they had hoped, and instead increased tariffs and trade restrictions. For example, recent tariff hikes by New Delhi on cell phones and other products have elevated longstanding U.S. concerns about India's tariff regime, and President Trump has called India the "tariff king."[29] Other U.S. concerns include inadequate intellectual property protection and enforcement, and restrictive new rules on e-commerce and localization of certain financial data flows—which affect major U.S. companies, such as Amazon, Walmart-owned Flipkart, Visa, and MasterCard. The United States and India also often have opposing stances on multilateral trade issues in the World Trade Organization.

---

[26] This section written by Shayerah Ilias Akhtar, Specialist in International Trade and Finance.

[27] See also CRS In Focus IF10384, *U.S.-India Trade Relations*, by Shayerah Ilias Akhtar and K. Alan Kronstadt.

[28] Some analysts view the prime minister's strong political mandate as offering a new opportunity for the government to undertake difficult economic reforms. Modi may choose to initiate major development projects rather than bold reforms, and the possibility that economic nationalism might drive his policies leads to concerns about further deterioration of trade links with the United States (Alyssa Ayres, "India's Economy Needs Tougher Reform—How Will Modi Use His Election Mandate?," *Time*, May 25, 2019). Some observers reject the notion that Modi is a genuine economic "reformer" (Sadanand Dhume, "BJP's Unsustainable Muslim Bashing" (op-ed), *Wall Street Journal*, April 5, 2019; Gurcharan Das, "The Modi Mirage," *Foreign Affairs*, April 11, 2019).

[29] The White House, "Remarks by President Trump on the United States-Mexico-Canada Agreement," press release, October 1, 2018.

India's Economy Under Modi

Economists and investors appear to view Modi's reelection as a net positive for India's economy, especially by providing a level of stability that a coalition government could not. Controlling inflation and reducing corruption are seen by many analysts as areas of success.[30] Implementing a uniform sales tax and enacting a new bankruptcy code likely contributed to India's rapid rise in the World Bank's annual *Doing Business* indices (in 2014, India was ranked 134th of 189 countries; by 2019, it had risen to 77th).[31] India's stock market spiked in value by 2% to a record high upon news of the BJP/Modi mandate. Other notable initiatives include electrification, building numerous new roads and toilets, providing affordable cooking gas for the poor, and a major health insurance initiative. According to the Economist Intelligence Unit, Modi's messaging has resonated with India's aspiring middle classes and his government appears to have maintained the support of business interests.[32]

Yet India's economy is widely seen to be underperforming, and the 2016 demonetization effort was deemed broadly harmful for India's poor.[33] Allegedly insufficient attention to the rural economy fueled anti-government protests by farmers and, since 2014, sluggish job growth, falling wages, and slow and poorly implemented institutional reforms have elicited significant criticism.[34] When official employment data were finally leaked to the public after years under wraps, they painted an alarming picture, with unemployment in 2018 at a 45-year high of 6.1%, triple the level in 2012 when the Statistics Ministry survey was last conducted. A shortage of jobs for India's booming population may yet transform the country's "demographic dividend" into a "demographic debacle."[35]

---

[30] Berlin-based Transparency International's Corruption Perceptions Index 2018, which ranks countries by their perceived levels of public sector corruption, listed India 78th of 180 countries (the same as Turkey, among others), noting slow, but steady improvement in recent years (see https://www.transparency.org/cpi2018).

[31] Gurcharan Das, "The Modi Mirage," *Foreign Affairs*, April 11, 2019; see http://www.doingbusiness.org/en/reports/global-reports/doing-business-2019.

[32] Modi's India," BBC News, April 10, 2019; see the EIU narrative at http://www.eiu.com/topic/democracy-index.

[33] Demonetization came with the government's surprise November 2016 announcement that about 86% of all paper circulating paper currency at the time would cease to be legal tender. Two new replacement notes were issued slowly, causing country-wide economic shocks (stock indices dropped by 6%) and leaving millions of people standing in long ATM lines to collect the new tender. The approach was widely panned by Indian and international economists, and was identified as a central cause of economic slowdown thereafter. Still, surveys indicated that Indians broadly approved of the effort, despite the hardship that resulted (see Bhaskar Chakravorti, "Early Lessons from India's Demonetization Experiment," *Harvard Business Review*, March 14, 2017).

[34] "India Elections 2019: Farmers Are Hurting and Millions Are Voting," *New York Times*, April 20, 2019.

[35] "Unemployment Rate at Four-Decade High of 6.1% in 2017/18: NSSO Survey," *Business Standard* (New Delhi), January 30, 2019; "Nobel-Winner Paul Krugman Warns India Story Could End with Mass Unemployment," *Economic Times* (Mumbai), March 17, 2018.

The outlook for bilateral trade relations is unclear. Both sides have taken decisive actions on simmering issues. President Trump terminated India's eligibility for the Generalized System of Preferences (GSP),[36] effective on June 5, 2019, after determining that "India has not assured the United States that India will provide equitable and reasonable access to its markets."[37] This decision followed a U.S. investigation into India's market access practices and petitions by U.S. dairy and medical technology industries. In 2018, India was the largest beneficiary of GSP, with over one-tenth ($6.3 billion) of U.S. goods imports from India entering duty-free under the program. India, which called the eligibility termination "unfortunate," announced soon after that it would impose higher retaliatory tariffs on 28 U.S. products, including almonds, apples, and walnuts, in response to U.S. Section 232 (national-security-based) steel and aluminum tariffs.[38] During 2018, India repeatedly delayed applying retaliatory tariffs, in hopes of negotiating a resolution of bilateral trade issues. In late June, President Trump called India's 2019 imposition of retaliatory tariffs "unacceptable" and said they "must be withdrawn."[39]

At the June 2019 G-20 Summit, the two sides appeared to strike a more conciliatory tone. President Trump said a "very big" trade deal would be coming with India.[40] According to India's foreign secretary, the "trade ministers of both countries would meet at an early date and would try to sort out these issues."[41] Yet lack of progress in recent months reportedly is prompting the Administration to consider launching a Section 301 investigation of India's trade practices—this would make India the focus of

---

36 GSP is a U.S. trade program that confers duty-free tariff treatment to certain U.S. imports from eligible developing countries to support their economic development, as long as they meet eligibility requirements. See CRS Report RL33663, *Generalized System of Preferences (GSP): Overview and Issues for Congress*, by Vivian C. Jones.

37 The White House, "Proclamation to Modify the List of Beneficiary Developing Countries Under the Trade Act of 1974," May 31, 2019.

38 CRS Report R45249, *Section 232 Investigations: Overview and Issues for Congress*, coordinated by Rachel F. Fefer.

39 Neha Dasgupta, "Trump Demands Withdrawal of India's 'Unacceptable' Tariff Hike," Reuters, June 27, 2019.

40 "Trump Changes His Tone on Trade at G-20," *The Wall Street Journal*, June 28, 2019.

41 Government of India, Ministry of External Affairs, "Transcript of Media Briefing by Foreign Secretary after JapanUS-India Trilateral Meeting on the Sidelines of G20 Summit (June 28, 2019)," June 28, 2019.

the next major, in-depth investigation of unfair trade practices after China.[42] On one hand, Section 301 could be a new way to address long-standing issues of U.S. concern with respect to India. On the other hand, it could raise the risk of protracted bilateral trade tensions and tit-for-tat escalation of tariffs across many economic sectors.

Defense and Security Relations

Continuity in India's leadership may lead to continued rapid development of U.S.-India security cooperation, with U.S. leaders hoping that increased Indian capabilities will provide greater net security regionally and worldwide (in spite of some U.S. concerns about New Delhi's ties with Moscow). President Obama and Congress recognized India as a "major defense partner" (MDP) in 2016, a unique designation allowing India to receive license-free access to dual-use American technologies.[43] The MDP designation was created in large part to carry over a presumption of license approvals into the new U.S. administration. It was linked to India's joining the four major multilateral export control regimes to become eligible to receive licensing for the most advanced defense systems the United States exports.[44] In mid-2018, India received Strategic Trade Authorization Tier-1 status, putting it on par with NATO allies, and the two countries concluded a long-sought Communications, Compatibility, and Security Agreement (COMCASA), a militaryto-military "enabling agreement." The two governments also seek expanded collaboration via the Defense Technology and Trade Initiative, launched in 2012, which aims to facilitate greater defense trade and technology

[42] Adam Behsudi, "India Could Be Next on Trump's Investigate-and-Tariff Hit List," *POLITICO*, June 3, 2019. For background, see CRS In Focus IF10708, *Enforcing U.S. Trade Laws: Section 301 and China*, by Wayne M. Morrison.

[43] India's bespoke MDP status was formalized by the 114th Congress, conferred in Section 1292 of the National Defense Authorization Act for FY2017 (P.L. 114-328), which discusses "enhancing defense and security cooperation with India." See also the June 7, 2016, U.S.-India Joint Statement at http://go.usa.gov/x8EFV).

[44] CRS interviews with Defense Department officials, April 2018.

sharing.[45] India participates in formalized "quadrilateral consultations" with the United States, Japan, and Australia while downplaying prospects that the "Quad" may become a security-related architecture.

Defense trade is a leading facet of the bilateral partnership. India is now a major purchaser in the global arms market and a lucrative potential customer for U.S. companies. The two nations have signed defense contracts worth about $15 billion since 2008, up from $500 million in all previous years combined.[46] Washington seeks to identify sales that can proceed under the technology-sharing and co-production model sought by New Delhi while also urging reform in India's defense offsets policy.[47] Since 2002, the United States and India have held a series of increasingly complex combined bilateral exercises involving all military services—India now conducts more exercises and personnel exchanges with the United States than with any other country. A first-ever tri-service exercise is set for later in 2019.[48] Bilateral intelligence and counterterrorism cooperation has accelerated over the past decade. In addition to intelligence sharing, homeland security cooperation has included growing engagement between respective law enforcement agencies, especially in the areas of mutual legal assistance and extradition, and on cyberterrorism and cybersecurity. Terrorist groups operating from Pakistani territory are of special interest.

India broadly endorses the FOIP strategy pursued by Washington, and it benefits from the higher visibility this strategy provides for India's global role and for its immediate region. Yet India has not fully relinquished the "nonalignment" posture it maintained for most of the Cold War (more recently pursuing "strategic autonomy" or a "pragmatic and outcome-

---

[45] See https://www.state.gov/u-s-security-cooperation-with-india.

[46] Notable recent deals include those for *Seahawk* maritime and *Apache* attack helicopters ($2.6 billion and $2.3 billion, respectively), *Poseidon* maritime patrol aircraft ($3 billion), and M777 towed howitzers ($737 million), among others (see the State Department's June 4, 2019, fact sheet at https://www.state.gov/u-s-security-cooperation-with-india).

[47] Since 2005, India has required that 30% of any defense deal valued at more than Rs3 billion (about $50 million) must be reinvested in the Indian economy, a requirement that many firms find difficult to meet.

[48] According to the Pentagon, joint exercises facilitate "an effective combined response to military missions such as providing humanitarian assistance and disaster relief, countering piracy, and responding to transnational threats and terrorism (see https://www.hsdl.org/?view&did=802816).

oriented foreign policy"[49]). Thus, Modi has articulated a vision of a free, open, and *inclusive* Indo-Pacific, and India remains wary of joining any nascent or potential security architectures that could antagonize Beijing.[50]

India's Other Foreign Relations

India-China

Modi's win likely means a continuation of New Delhi's multilateralist/multipolar approach to international politics in Asia, as well as efforts to resist Chinese "assertiveness" in South Asia.[51] India's relations with China have been fraught for decades, with signs of increasing enmity in recent years. Areas of contention include major border and territorial disputes, China's role as Pakistan's primary international benefactor, the presence in India of the Dalai Lama and a self-described Tibetan "government," and China's growing presence in the Indian Ocean region, which many Indians view as an encroachment in their neighborhood. New Delhi is ever watchful for signs that Beijing seeks to "contain" Indian influence both regionally and globally. China's BRI—with "flagship" projects in Pakistan—is taken by many in India (and elsewhere) as an expression of Beijing's hegemonic intentions.

India-Pakistan

The United States has long sought to assist in reducing India-Pakistan conflict and its impact on developments in Afghanistan. A surge of Indian nationalism grew out of a February 2019 international crisis involving

---

[49] Ministry of External Affairs Annual Report 2017-2018.

[50] By some accounts, India is poorly suited to serve as the western anchor of the FOIP, given its apparent intentions to maintain strategic autonomy, and its alleged lack of the will and/or the capacity to effectively counterbalance China. Moreover, many in India consider a FOIP conception that terminates at India's western coast to be "a decidedly U.S.- centric, non-Indian perspective" that omits a huge swath of India's strategic vista to the west (Michael Swaine, "Creating an Unstable Asia: The U.S. 'Free and Open Indo-Pacific' Strategy," Carnegie Endowment for International Peace, March 2, 2018).

[51] "China Says Ties with India Insulated from Differences on Belt and Road," Hindu (Chennai), April 19, 2019. See also CRS Report R45194, *China-India Great Power Competition in the Indian Ocean Region: Issues for Congress*, by Bruce Vaughn.

Pakistan. Coming just weeks before voting began in India, the confrontation was widely seen to have boosted Modi's electoral prospects.[52] In contrast to his 2014 inauguration, when Pakistan's then-prime minister was an invited guest, Modi in 2019 omitted Pakistani Prime Minister Imran Khan from swearing-in festivities. It is unclear if the Indian leader will seek to use his political capital to launch a new peace initiative with Islamabad or will continue pursuing punitive policies that aim to isolate Pakistan internationally. Some analysts contend that Modi now has sufficient standing to change tack. Others, however, suggest that his reelection "will be projected as a vindication of his belligerent policy toward Pakistan."[53]

India-Russia

India maintained close ties with Russia throughout much of the Cold War and continues to rely on Moscow for the bulk of its defense imports. With the 2017 enactment of the Countering America's Adversaries Through Sanctions Act (CAATSA, P.L. 115-44) in U.S. law, India's continued major arms purchases from Russia—most prominently a current multi-billiondollar deal to purchase the Russian-made S-400 air defense system—could trigger U.S. sanctions. Although Congress subsequently provided for a national security waiver of these sanctions, the Administration has consistently counseled India to cancel the purchase and consider U.S.- supplied alternatives, while New Delhi insists that it will go forward in pursuit of its own national interest, a position possibly hardened with the BJP mandate.[54]

---

[52] In February, a suicide car bombing in the Kashmir Valley killed 40 Indian paramilitary troops and was blamed on a Pakistan-based terrorist group; a subsequent retaliatory Indian airstrike on Pakistani territory and a brief jet aircraft battle elicited global concern about escalation, but the crisis ended without further major conflict ("Pakistan Standoff Helps India's Modi Shift Focus from Jobs," Associated Press, March 15, 2019; "Missiles Maketh the Man," *Economist* (London), May 4, 2019).

[53] "After Divisive India Election, Some See Moment for Dialogue with Pakistan," *Wall Street Journal*, May 28, 2019; Zakir Hussein, "If Modi Returns" (op-ed), *Dawn* (Karachi), May 22, 2019.

[54] Section 231 of CAATSA targets "significant transactions" with Russia's defense or intelligence sectors. Section 1294 of the National Defense Authorization Act for FY2019 (P.L. 115-232) permits the President to waive CAATSA sanctions, but the conditions are fairly stringent: waiver authority requires the President to certify that a waiver is in the U.S.

India-Iran

India has historically friendly relations with Iran, a country that lately has supplied about 30% of India's energy imports. It also opposes any potential acquisition of nuclear weapons by Iran and supports the Joint Comprehensive Plan of Action. Historically averse to unilateral (non-U.N.) sanctions, New Delhi until recently enjoyed exemption from U.S. efforts targeting Iran's energy sector. In April 2019, the Trump Administration ended such exemptions, and New Delhi has issued conflicting statements about its cessation of Iranian oil purchases while informing Washington that such cessation "comes at a cost."[55] Continued tensions in the Persian Gulf and/or a longer-term boycott of Iranian oil could be disruptive to the Indian economy; Modi's strong mandate could place limits on his willingness to abide such disruption.

Other Notable Relations

The Trump Administration's South Asia strategy has included calls for greater Indian involvement in Afghanistan, even as such engagement vexes Pakistan (New Delhi has committed about $3 billion to Afghan reconstruction to date). Islamabad is wary of the Indian presence in Afghanistan and accuses New Delhi of supporting anti-Pakistan groups there, a dynamic that can in turn affect U.S. efforts to sustain Pakistan's help in facilitating Afghanistan reconciliation. Many analysts expect India-Afghanistan ties to grow stronger with Modi's reelection, and New Delhi appears wary of any precipitous U.S. withdrawal from Afghanistan. Meanwhile, India's deepening "strategic partnership" with Japan is a major aspect of New Delhi's broader "Act East" policy and a key axis in the greater FOIP strategies broadly pursued by all three governments participating in a newly established U.S.-Japan-India Trilateral Dialogue. Prime Minister Modi appears to have a convivial personal relationship with

---

national security interest and that a government offered a waiver is significantly reducing the proportion of its total defense equipment produced by Russia, among other provisions.

[55] "India Says Iran Oil Imports Not Stopped," *Hindu* (Chennai), July 3, 2019. See also CRS Report RS 20871, *Iran Sanctions*, by Kenneth A. Katzman.

his Japanese counterpart and bilateral ties are seen as likely to strengthen going forward.

Human Rights Issues

While Prime Minister Modi and his party have long sought to emphasize development and good governance, the 2019 election cycle revolved around nationalism and other emotive issues, with many observers arguing that Hindu majoritarianism is a threat both to India's religious minorities and to the country's syncretic traditions. According to the U.S. State Department and independent watchdogs, India is the site of numerous human rights violations, many of them serious and some perpetrated, or at least tolerated, by state actors.[56] Many observers are concerned about the impact of growing religious bigotry and Hindu nationalism on human rights.[57] The BJP is an openly Hindu nationalist party and Prime Minister Modi is a self-avowed Hindu nationalist.[58] In 2005, Modi was denied a U.S. visa over concerns about his role in government during lethal anti-Muslim violence in 2002.[59] Modi hit conciliatory notes in a national address three days after the 2019 election results were announced, vowing to seek the trust of minority groups and to work for the good of all Indians.

---

56 "A lack of accountability for misconduct persisted at all levels of government, contributing to widespread impunity" (see the State Department's *2018 Country Reports on Human Rights Practices: India* at https://go.usa.gov/xy2W4).

57 See CRS Report R45303, *India: Religious Freedom Issues*, by K. Alan Kronstadt.

58 The BJP was born in 1980 as the political wing of a hardline Hindu nationalist and social service group, the Rashtriya Swayamsevak Sangh (RSS or "National Volunteer Organization"); 99% of BJP seats are held by Hindus, versus about 71% of Congress seats ("What You Need to Know About India's BJP," Al Jazeera (Doha), May 23, 2019).

59 In 2002, on Modi's watch, days-long communal rioting left at least 1,000 people dead, most of them Muslims targeted by Hindu mobs. The State Department later denied Modi a U.S. visa in 2005 under the 1998 International Religious Freedom Act, which allows for barring entry to foreign government officials found to be complicit in severe violations of religious freedom. The U.S. government subsequently had no official contacts with Modi until late 2013.

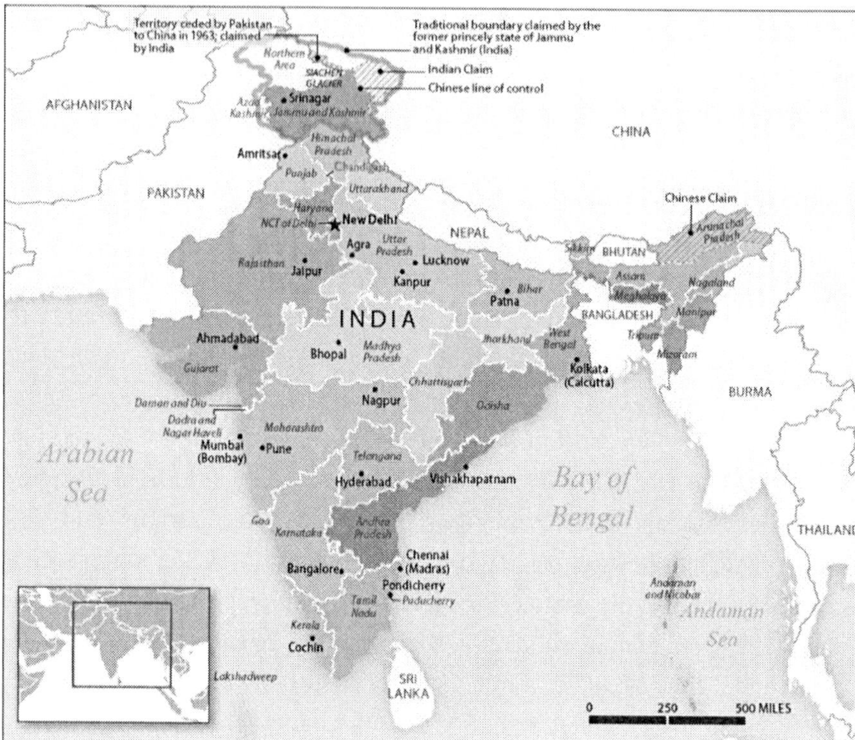

Source: CRS in consultation with the Department of State (2016); Department of State international boundary files (2015); Esri (2014); and DeLorme (2014).

Note: Limits shown do not reflect U.S. government policy on boundary representation or sovereignty.

Figure 1. Map of Indian States.

Human rights nongovernmental organizations and social service groups have seen their Indian operations constrained in recent years, and observers are watching closely for signs that the Modi/BJP mandate will lead to renewed efforts toward Hindu nationalist goals.[60] Some of these—including laws preventing religious conversions and cow slaughter—continue to cause sparks in U.S.-India relations, including explicit BJP

---

[60] See the State Department's *2018 Report on International Religious Freedom* at https://go.usa.gov/xy2WM and USCIRF reports at https://go.usa.gov/xy2Wz. In 2017, a bipartisan group of five U.S. Senators penned a letter urging President Donald Trump to raise the issue of India's deteriorating religious freedom during Prime Minister Modi's mid-year visit to Washington, DC (see the June 23, 2017, letter at https://go.usa.gov/xUAGG).

criticism of the U.S. government for alleged "bias" against Modi.[61] Future moves by the Modi government on other "Hindutva" policies could increase national divisions and lead to further international opprobrium.[62]

---

[61] "BJP Slams U.S. for 'Bias' Against PM Modi," *Hindu* (Chennai), June 23, 2019.

[62] Such policy goals include scaling back laws and social programs designed to assist India's religious minority communities, establishing a uniform civil code, repealing Article 370 of the Indian Constitution (which grants limited autonomy to the state of Jammu & Kashmir), and constructing a Ram temple on the site of the razed Babri Mosque, among others ("India's New Government Signals Hindu and India-First Goals," Associated Press, May 31, 2019).

In: Key Congressional Reports for June 2019 ISBN: 978-1-53616-666-8
Editor: Tara T. Robertson

Chapter 3

NORTH KOREA: LEGISLATIVE BASIS FOR U.S. ECONOMIC SANCTIONS (UPDATED)*

Dianne E. Rennack

ABSTRACT

U.S. economic sanctions imposed on North Korea are instigated by that country's activities related to weapons proliferation, especially its tests since 2006 of nuclear weapons and missile technology; regional disruptions; terrorism; narcotics trafficking; undemocratic governance; and illicit activities in international markets, including money laundering, counterfeiting of goods and currency, and bulk cash smuggling. The sanctions have the following consequences for U.S.- North Korea relations:

- Trade is limited to food, medicine, and other humanitarian-related goods, all of which require a license. Imports from North Korea are prohibited as of June 2011; exports to North Korea of most U.S.-origin goods, services, or technology are prohibited as of March 2016. Trade in luxury goods is banned. The Department of Commerce

* This is an edited, reformatted and augmented version of Congressional Research Service, Publication No. R41438, dated June 6, 2019.

denies export licenses for reasons of nuclear proliferation, missile technology, U.N. Security Council requirements, and international terrorism.

- Arms sales and arms transfers are fully denied.
- Financial transactions are prohibited. U.S. persons are prohibited from providing financial services for the purpose of evading sanctions, or from providing financial services to a person or entity designated for sanctions. The President, in September 2017, authorized the Secretary of the Treasury to designate for sanctions any foreign financial institution that conducts or facilitates "any significant transaction on behalf of any [designated] person," or "in connection with trade with North Korea."
- North Korea is designated as a jurisdiction of primary money laundering concern by the Department of the Treasury's Financial Crimes Enforcement Network (FinCEN), effective December 9, 2016.
- U.S. new investment is prohibited, and investment in North Korea's transportation, mining, energy, or financial sectors is prohibited. North Korea is also ineligible to participate in any U.S. government program that makes credit, credit guarantees, or investment guarantees available.
- U.S. foreign aid is minimal and mostly limited to refugees fleeing North Korea; broadcasting into the country; nongovernmental organization programs dedicated to democracy promotion, human rights, and governance; and emergency food aid. In past years, aid related to disabling and dismantling the country's nuclear weapons program has been made available. By law, U.S. representatives in the international financial institutions (IFI) are required to vote against any support for North Korea due to its nuclear weapons ambitions and international terrorism. Human rights and environmental activities would also likely result in U.S. objections to North Korea's participation in the IFI.
- U.S.-based assets are blocked for North Korean individuals, entities, aircraft, and vessels designated by the Department of the Treasury's Office of Foreign Assets Control (OFAC). U.S. persons are prohibited from entering into trade and transaction with these designees and, foreign financial institutions could become subject to U.S. sanctions for facilitating transactions for designated persons.
- Kim Jong-un, the Korean Workers' Party, and others—banks, shipping companies, seagoing vessels, state agencies, and other individuals affiliated with the state's security regime—are identified as being among those engaged in illicit and punishable activities, possibly including nuclear or ballistic missile programs, undermining cybersecurity, censorship, and sanctions evasion. As a result,

effective March 15, 2016, any of their assets under U.S. jurisdiction are frozen, and U.S. persons and entities are prohibited from entering into trade and transactions with the designees.

- U.S. travel requires a special validation passport issued by the State Department. Such passports are reserved only for travel in the U.S. national interest and are intended for professional reporters, officials with the American Red Cross or International Committee of the Red Cross, or those who have a "compelling humanitarian" justification.

From the outbreak of the Korean War in 1950, the United States had imposed fairly comprehensive economic, diplomatic, and political restrictions on North Korea. In 1999, however, President Clinton announced the United States would lift many restrictions on U.S. exports to and imports from North Korea in areas other than those controlled for national security concerns; the Departments of Commerce, Treasury, and Transportation issued new regulations a year later that implemented the new policy. On June 26, 2008, President George W. Bush delisted North Korea as a state sponsor of international terrorism, and removed restrictions based on authorities in the Trading with the Enemy Act and the terrorism designation, replacing them with more circumscribed economic restrictions related to proliferation concerns.

The U.S. sanctions are a result of requirements incorporated into U.S. law by Congress, decisions made in the executive branch to exercise discretionary authorities, and obligations placed on member states of the United Nations by the U.N. Security Council. Though the President, in accordance with the Constitution, leads the way in conducting foreign policy, Congress holds substantial power to shape foreign policy by authorizing and funding programs, advising on appointments, and specifically defining the terms of engagement in accordance with U.S. political and strategic interests. This chapter presents the legislative basis for U.S. sanctions policy toward North Korea. These sanctions are a critical tenet of the larger bilateral relationship, and this chapter highlights Congress's role and responsibility in determining the nature of U.S.-North Korea relations.

BACKGROUND

The United States imposes economic sanctions on North Korea (DPRK) for activities related to weapons proliferation, particularly its pursuit of nuclear weapons capability and a missile delivery system; regional disruptions; narcotics trafficking; undemocratic governance; and illicit activities in international markets, including money laundering, counterfeiting of goods and currency, and bulk cash smuggling. The entirety of North Korea is identified by the United States as a "jurisdiction of money laundering concern," which makes DPRK-origin or DPRK-related transactions to or through the U.S. financial system off limits. And the Government of North Korea is designated as a state sponsor of acts of international terrorism, which quashes most U.S. foreign aid, trade, investment, and support in international financial institutions.

United States law has been applied to North Korea in the following ways in response to the North Korean government's objectionable activities:[1]

- North Korea poses a threat to U.S. national security because of "the current existence and risk of the proliferation of weapons-usable fissile material on the Korean Peninsula," as declared by President George W. Bush on June 26, 2008, under the terms of the National Emergencies Act (NEA) and the International Emergency Economic Powers Act (IEEPA);[2]
- North Korea is cited by the United Nations Security Council[3] for its nuclear weapons and ballistic missiles pursuits, withdrawal

---

[1] Appendix A lists U.S. statutory authorities used to form the economic sanctions regime.

[2] Executive Order 13466, "Continuing Certain Restrictions With Respect to North Korea and North Korean Nationals," 73 F.R. 36787, June 26, 2008; 31 C.F.R. Part 510, November 4, 2010. The same day, the President found that continuing the national emergency first proclaimed under authority of the Trading With the Enemy Act (Presidential Proclamation 2914; December 16, 1950; 15 F.R. 9029) was "no longer in the national interest of the United States." Presidential Proclamation 8271; June 26, 2008; 73 F.R. 36785.

The President is required to continue annually any national emergency he issues under the National Emergencies Act, or it expires, along with the sanctions established under the International Emergency Economic Powers Act (IEEPA). President Obama renewed the national emergency declared in Executive Order 13466 in memoranda issued on June 24, 2009 (74 F.R. 30457), and annually since then, most recently on June 22, 2018 (83 F.R. 29661).

[3] Appendix B compares U.N. Security Council resolutions with steps taken by the United States, in both legislation and executive orders. United Nations Security Council Resolution 1718 (2006), October 14, 2006; UNSC Resolution 1874 (2009), June 12, 2009; UNSC Resolution 2087 (2013), January 22, 2013; UNSC Resolution 2094 (2013), March 7, 2013; UNSC Resolution 2270 (2016), March 2, 2016; UNSC Resolution 2321 (2016), November 30, 2016; UNSC Resolution 2356 (2017), June 2, 2017; UNSC Resolution 2371 (2017), August

from the Treaty on the Non-Proliferation of Nuclear Weapons, and contribution to regional tensions; the United States meets the requirements as a member state of the United Nations pursuant to the United Nations Participation Act of 1945 to implement sanctions adopted by the U.N. Security Council;

- North Korea committed an unprovoked attack that resulted in the sinking of a South Korean naval vessel, *Cheonan*; announced a new nuclear test and missile launches in 2009; and engaged in money laundering, counterfeiting of goods and currency, bulk cash smuggling, and narcotics trafficking, all in violation of U.N. Security Council Resolutions, leading President Barack Obama to expand the national emergency in 2010, 2011, 2015, and 2016. President Donald Trump, in 2017, cited the "provocative, destabilizing, and repressive actions and policies of the Government of North Korea"; its intercontinental ballistic missile launches; its nuclear tests; violation of Security Council resolutions; and human rights abuses to continue the national emergency and impose additional sanctions.[4]
- North Korea is a Marxist-Leninist state, with a Communist government, and stated as such in the Export-Import Bank Act of 1945, as amended, and further restricted under the Foreign Assistance Act of 1961;
- North Korea has engaged in proliferation of weapons of mass destruction, the State Department finds pursuant to the Arms Export Control Act; Export Administration Act of 1979; Iran,

---

5, 2017; UNSC Resolution 2375 (2017), September 11, 2017; and UNSC Resolution 2397 (2017), December 22, 2017.

[4] Executive Order 13551 of August 30, 2010, "Blocking Property of Certain Persons With Respect to North Korea," 75 F.R. 53837, September 1, 2010; Executive Order 13570 of April 18, 2011, "Prohibiting Certain Transactions With Respect to North Korea," 76 F.R. 22291, April 20, 2011; Executive Order 13687 of January 2, 2015, "Imposing Additional Sanctions With Respect to North Korea," 80 F.R. 819, January 6, 2015; Executive Order 13722 of March 15, 2016, "Blocking Property of the Government of North Korea and the Workers' Party of Korea, and Prohibiting Certain Transactions With Respect to North Korea," 81 F.R. 14943, March 18, 2016; and Executive Order 13810 of September 20, 2017, "Imposing Additional Sanctions With Respect to North Korea," 82 F.R. 44705. Both Presidents cited the United Nations Participation Act of 1945 (P.L. 79-264; 22 U.S.C. 287c) authorities in addition to those provided in NEA and IEEPA.

North Korea, and Syria Nonproliferation Act of 2000; and a national emergency declared by President George H. W. Bush relating to the proliferation of such weapons;

- North Korea is not cooperating fully with U.S. antiterrorism efforts, the State Department has determined, under terms of the Arms Export Control Act;
- North Korea has detonated a nuclear explosive device, President George W. Bush has determined,[5] pursuant to the Arms Export Control Act, the Atomic Energy Act, and the Export-Import Bank Act of 1945;
- North Korea engaged in "provocative, destabilizing, and repressive actions and policies," including "destructive, coercive cyber-related actions during November and December 2014," actions in violation of a multitude of U.N. Security Council resolutions, and commission of serious human rights abuses;[6] and
- North Korea "used chemical weapons in violation of international law or lethal chemical weapons against its own nationals." On February 22, 2018, the Secretary of State determined that the government of North Korea was responsible for the lethal 2017 nerve agent attack on Kim Jong Nam, the half-brother of North Korean leader Kim Jong-un, in Malaysia. Sanctions went into effect on March 5, 2018, though they are largely redundant with restrictions already in place.[7]

---

[5] On December 7, 2006, President Bush determined that North Korea, a non-nuclear-weapon state, had detonated a nuclear explosive device, citing Section 102(b) of the Arms Export Control Act and Section 129 of the Atomic Energy Act. Presidential Determination No. 2007-07. *Public Papers of the President*. December 18, 2006.

[6] Executive Order 13687 of January 2, 2015, "Imposing Additional Sanctions With Respect to North Korea, 80 F.R. 819, January 6, 2015. The Executive Order also draws on authorities granted the President in the Immigration and Nationality Act (8 U.S.C. 1182(f)) to deny entry into the United States of any person designated pursuant to IEEPA authorities. See Presidential Proclamation 8693 of July 24, 2011, "Suspension of Aliens Subject to United Nations Security Council Travel Bans and International Emergency Economic Powers Act Sanctions," 76 F.R. 44751.

[7] Department of State Public Notice 10340, "Bureau of International Security and Nonproliferation; Determinations Regarding Use of Chemical Weapons by North Korea Under the Chemical and Biological Weapons Control and Warfare Elimination Act of 1991," 83 F.R. 9362, March 5, 2018.

At the President's discretion, North Korea also could be subject to economic sanctions provided in three provisions of law addressing human rights conditions: the Foreign Assistance Act of 1961, the International Religious Freedom Act of 1998,[8] and the Trafficking Victims Protection Act of 2000.[9] Sanctions under these acts would be largely redundant with restrictions already in place.

The United States' concerns about North Korea's pursuit of nuclear weapons capability emerged in the 1980s when that country's nuclear weapons program became apparent.[10] In the 1990s, the two countries negotiated and signed (in 1994) an Agreed Framework to freeze North Korea's plutonium-based nuclear energy program and provide heavy fuel oil until light-water reactors could be brought online, all funded to varying degrees by the European Union, Japan, South Korea, and the United States. Through the late 1990s, the United States and North Korea engaged in negotiations, initially to curtail missile tests and expanded to include differences on nuclear weapons and international terrorism. In October 2002, it came to light in negotiations between U.S. and North Korean government officials that North Korea was pursuing a uranium-based nuclear weapons capability. Diplomacy over North Korea's nuclear weapons program then entered a new phase; the Agreed Framework was

---

[8] Under the International Religious Freedom Act of 1998 (P.L. 105-292; 22 U.S.C. 6401 et seq.), North Korea is identified as a Tier 1 country of particular concern (CPC), the worst ranking available. The U.S. Commission on International Religious Freedom recommended, in part, that the United States should "[m]ake any future discussions regarding the easing or lifting of sanctions contingent upon North Korea's sincere and demonstrable efforts to improve its religious freedom and human rights record consistent with international standards." U.S. Commission on International Religious Freedom, *2019 Annual Report*, April 2019, pp. 66-71.

[9] Under the Trafficking Victims Protection Act of 2000 (Division A of P.L. 106-386; 22 U.S.C. 7101 et seq.), the government of the DPRK is identified as a Tier 3 government, the worst ranking available, for its persistent failure "to meet the minimum standards for the elimination of trafficking [in persons] and ... not making efforts to do so." The DPRK government is noted for its "continued state-sponsored human trafficking through its use of forced labor in prison camps, as part of an established system of political repression, and in labor training centers, facilitation of forced labor of students, and its exportation of forced labor to foreign companies." Department of State. Trafficking in Persons Report, June 2018, pp. 255-256.

[10] The Arms Control Association maintains a comprehensive *Chronology of U.S.-North Korean Nuclear and Missile Diplomacy*, available at https://www.armscontrol.org/factsheets/dprkchron.

abandoned and the United States, North Korea, South Korea, China, Japan, and Russia convened a new forum—the Six Party Talks, which held its first negotiations in August 2003. Despite several steps forward, including the United States ending decades-long sanctions imposed at the outset of the 1950-1953 conflict and its delisting of North Korea as a supporter of international terrorism, the Six Party Talks collapsed in late 2008.

During this period and through 2017, North Korea has engaged in a number of acts that the international community deems provocative. It has tested short-, mid-, and long-range ballistic missiles since mid-2006, with varying degrees of success, abrogating a moratorium it complied with since 1998. In early 2009, North Korea began to test satellite launches, contrary to U.N. Security Council restrictions, and may have succeeded in placing a satellite in orbit in 2012. Most recently, it has tested missile launches from under the sea.

On October 9, 2006, May 25, 2009, February 12, 2013, January 6, 2016, September 9, 2016, and again on September 3, 2017, North Korea reported that it had detonated a nuclear explosive device. The international community responded to the missile tests and nuclear detonations by taking the issue to the U.N. Security Council, which adopted resolutions that condemned the weapons tests and called on member states to impose economic sanctions.

On March 26, 2010, a South Korean Navy ship, the *Cheonan*, was struck by a torpedo while sailing in the West Sea. The ship sank and 46 crew members were killed. The South Korean Ministry of National Defense formed a Civilian-Military Joint Investigation Group—with participants from five other nations including the United States—which found that the *Cheonan* was torpedoed by a North Korean submarine. North Korea denied involvement. The U.N. Security Council, in a presidential statement, condemned the attack, and acknowledged both the findings of the Investigation Group and the disavowal by North Korea.[11]

[11] Civilian-Military Joint Investigation Group, *On the Attack Against the ROK Ship Cheonan*, Ministry of National Defense, Republic of Korea, September 2010; U.N. Security Council, Presidential Statement (U.N. document, S/PRST/2010/13, July 9, 2010).

President Obama assigned responsibility to North Korea for the sinking of the *Cheonan* when, on August 30, 2010, he announced he was expanding the scope of the national emergency declared in 2008, and the United States was taking additional steps to curtail economic activity with North Korea:

> the continued actions and policies of the Government of North Korea, manifested most recently by its unprovoked attack that resulted in the sinking of the Republic of Korea Navy ship *Cheonan* and the deaths of 46 sailors in March 2010; its announced test of a nuclear device and its missile launches in 2009; its actions in violation of UNSCRs 1718 and 1874, including the procurement of luxury goods; and its illicit and deceptive activities in international markets through which it obtains financial and other support, including money laundering, the counterfeiting of goods and currency, bulk cash smuggling, and narcotics trafficking, destabilize the Korean peninsula and imperil U.S. Armed Forces, allies, and trading partners in the region, and thereby constitute an unusual and extraordinary threat to the national security, foreign policy, and economy of the United States.[12]

President Obama's explicit identification of all features of North Korea's objectionable behavior constituting the threat is unusual compared to other invocations of his IEEPA authorities.[13] The statute requires only that the President find that a threat to U.S. national security, foreign policy, or economy exists, and that its source is "in whole or substantial part outside the United States." The President identified North Korea's attack of the *Cheonan* and other acts of regional destabilization, pursuit of weapons of mass destruction and the means to deliver them, noncompliance with U.N. requirements, money laundering, counterfeiting, smuggling, and narcotics trafficking as compounding the threat. To this list, the President added, in January 2015, "cyber-related actions ... and commission of serious human rights abuses," the latter a nod to ground-

---

[12] Executive Order 13551, "Blocking Property of Certain Persons With Respect to North Korea," 75 F.R. 53837, September 1, 2010.

[13] 50 U.S.C. 1701 notes.

breaking efforts in December 2014 in the United Nations General Assembly to refer reports of human rights atrocities in North Korea's prison system to both the U.N. Security Council and the International Criminal Court. Justifying the declaration of the emergency based on these wide-ranging activities accomplishes a number of goals

- It confirms the United States' full support of and participation in implementation of the U.N. Security Council resolutions.
- It provides a clear list of concerns members of the U.S. diplomatic corps might raise and emphasize when speaking with North Korea's trading partners and benefactors.
- It states indisputable goals for North Korea to strive toward, meet, and surpass.

It should be noted, however, that each of these forms of objectionable behavior likely would be grounds, under current law, for restricting trade, aid, arms sales, and access to assets even if the national emergency were to be revoked.

U.S. Economic Sanctions Currently in Place

United States economic sanctions imposed on North Korea, as a result both of requirements in U.S. law and decisions made in the executive branch to exercise discretionary authorities, have the following impact:

- Trade is limited to food, medicine, and other humanitarian-related goods, all of which require a license. Imports from North Korea are prohibited as of June 2011; exports to North Korea of most U.S.-origin goods, services, or technology are prohibited as of March 2016. U.S. persons are further prohibited from engaging in transactions with the Government of North Korea or the Workers' Party of Korea related to trade in metal, graphite, coal, or software. Trade in luxury goods is banned. The Department of Commerce denies export licenses for reasons of nuclear proliferation, missile technology, U.N. Security Council requirements, and international terrorism. Using a North Korea-flagged vessel for any transaction is prohibited.

- Arms sales and arms transfers are fully denied.
- Financial transactions are prohibited. U.S. persons are prohibited from providing financial services for the purpose of evading sanctions, or from providing financial services to a person or entity designated for sanctions. The President, in September 2017, authorized the Secretary of the Treasury to designate for sanctions any foreign financial institution that conducts or facilitates "any significant transaction on behalf of any [designated] person," or "in connection with trade with North Korea."[14]
- North Korea is designated as a jurisdiction of primary money laundering concern by the Department of the Treasury's Financial Crimes Enforcement Network (FinCEN), effective December 9, 2016. FinCEN had issued advisories and warnings about North Korea's deceptive and illicit banking practices as early as 2005. This rule is subject to review by the U.S. financial sector and other departments of the U.S. government. If FinCEN's recommendations are adopted, U.S. banks will be prohibited from providing U.S. correspondent accounts to third-country banks to process transactions for North Korean financial institutions.
- U.S. new investment is prohibited, and investment in North Korea's transportation, mining, energy, or financial sectors is prohibited. North Korea is also ineligible to participate in any U.S. government program that makes credit, credit guarantees, or investment guarantees available.
- U.S. foreign aid is minimal and mostly limited to refugees fleeing North Korea; broadcasting into the country; nongovernmental organization programs dedicated to democracy promotion, human rights, and governance; and emergency food aid. In past years, aid

---

[14] Section 4, Executive Order 13810 of September 20, 2017; 82 F.R. 44706.

related to disabling and dismantling the country's nuclear weapons program has been made available. By law, U.S. representatives in the international financial institutions (IFI) are required to vote against any support for North Korea due to its nuclear weapons ambitions and international terrorism. Human rights and environmental activities would also likely result in U.S. objections to North Korea's participation in the IFI.

- U.S.-based assets are blocked for North Korean individuals, entities, aircraft, and vessels designated by the Department of the Treasury's Office of Foreign Assets Control (OFAC). U.S. persons are prohibited from entering into trade and transaction with these designees and, most recently, foreign financial institutions could become subject to U.S. sanctions for facilitating transactions for designated persons.
- Kim Jong-un, the Korean Workers' Party, and others—banks, shipping companies, seagoing vessels, state agencies, and other individuals affiliated with the state's security regime—are identified as being among those engaged in illicit and punishable activities, possibly including nuclear or ballistic missile programs, undermining cybersecurity, censorship, and sanctions evasion. As a result, effective March 15, 2016, any of their assets under U.S. jurisdiction are frozen, and U.S. persons and entities are prohibited from entering into trade and transactions with the designees.
- U.S. travel requires a special validation passport issued by the State Department. Such passports are reserved only for travel in the U.S. national interest and are intended for professional reporters, officials with the American Red Cross or International Committee of the Red Cross, or those who have a "compelling humanitarian" justification.[15]

[15] U.S. Department of State. Bureau of Consular Affairs, "Passport for Travel to North Korea," September 1, 2017 (expires March 31, 2021), https://travel.state.gov/content/travel/en/passports/requirements/passport-for-travel-to-northkorea.html.

The U.S. sanctions are a result of requirements incorporated into U.S. law by Congress, decisions made in the executive branch to exercise discretionary authorities, and obligations placed on the United States, as a member state of the United Nations, by the U.N. Security Council. Economic sanctions may restrict or prohibit all manner of bilateral relationships, but broadly can be categorized as impeding aid, trade, travel, and the finances related to these activities. Three Executive Branch Departments—State, Commerce, and Treasury—together have the lion's share of responsibilities to administer the restrictions affecting these activities.

Trade

The United States curtails trade with North Korea for reasons of regional stability, that country's support for acts of international terrorism, lack of cooperation with U.S. antiterrorism efforts, proliferation, and its status as a Communist country and a nonmarket economy. The United States also prohibits transactions relating to trade with certain North Korean entities identified as those who procure luxury goods, launder money, smuggle bulk cash, engage in counterfeiting goods and currency, and traffic in illicit narcotics.

National Emergency Because of Threat to U.S. National Security

Trade with North Korea is significantly restricted because of that country's demonstrated pursuit of nuclear weapons and the means to deliver them. In addition, the Department of Commerce has treated North Korea as a state sponsor of international terrorism, even in the period that the North Korean government was removed from the terrorism list, from mid-2008 through most of 2017.[16] Secretary of State Rex Tillerson redesignated the Government of North Korea as a state sponsor of international terrorism in late 2017.

---

[16] See, especially, 15 C.F.R. Part 742.19, "Anti-terrorism: North Korea."

Commerce's primary means of controlling exports is through the administration of Export Administration Regulations (EAR), in which goods to be controlled for a variety of reasons— national security, foreign policy, short supply, or compliance with international agreements, to name a few—are categorized. Recipient countries are also characterized, from allies for which little licensing is required, to rogue states for which export licensing is all but completely denied. Commerce identifies North Korea among the most restricted trade destinations—Country Group E:1, Terrorist Supporting Countries—which severely limits its access to computers, software, national security-controlled items, items on the Commerce Control List (CCL),[17] and service or repair of such items. A U.S. exporter intending to ship any goods subject to the Export Administration Regulations (EAR), except for food and medicine not on the CCL, is required to obtain an export license.

Commerce also identifies North Korea in the most restrictive country group—Country Group E. U.S. exports to countries in Group E are restricted for reasons of international terrorism.[18] As a result, U.S. exporters are likely to be denied licenses to export any controlled item if North Korea is the destination. In June 2007, Commerce eased licensing requirements so that food, medicine, and humanitarian assistance items could be made available, but at the same time imposed new licensing

---

[17] The Secretary of Commerce establishes and administers the Commerce Control List (CCL), goods controlled for national security reasons, pursuant originally to Section 5(c) of the Export Administration Act of 1979 (P.L. 96-72; 50 U.S.C. App. 2404(c)). Most of the 1979 Act, including section 5, was repealed with the enactment of the Export Controls Act of 2018 (Title XVII, Subtitle B, Part I, P.L. 115-232; 50 U.S.C. 4811 et seq.). Section 1768(a) of that Act (50 U.S.C. 4826) provides that:

All delegations, rules, regulations, orders, determinations, licenses, or other forms of administrative action that have been made, issued, conducted, or allowed to become effective under the Export Administration Act of 1979 (50 U.S.C. 4601 et seq.) (as in effect on the day before the enactment of this Act and as continued in effect pursuant to the International Emergency Economic Powers Act (50 U.S.C. 1701 et seq.)), or the Export Administration Regulations, and are in effect as of the date of the enactment of this Act [August 13, 2018], shall continue in effect according to their terms until modified, superseded, set aside, or revoked under the authority of this part.

[18] 15 C.F.R. Part 740 Supp. 1. Cuba, no longer designated as a state sponsor of international terrorism, is also designated on the E list. For arms embargoed country designations, see 22 C.F.R. Part 126.1.

prohibitions on the export and reexport of luxury goods to implement the terms of U.N. Security Council Resolution 1718.[19]

Thus, a U.S. company may apply for a license to export to North Korea, but for nearly all items other than food and medicine, there is a presumption of denial. The EAR identify license exceptions; those wishing to export to North Korea, however, are not eligible for these exceptions except in highly circumscribed instances.[20]

The Office of Foreign Assets Control, within the Department of the Treasury, must approve any U.S. importation from North Korea, and weighs all requests in the context of proliferation, money laundering, counterfeiting, bulk cash smuggling, narcotics trafficking, international terrorism, or other illicit economic activity, and who in North Korea might profit. Any transfer involving the Government of North Korea, any senior DPRK government official, or a DPRK person or entity designated as a Specially Designated National pursuant to any of the series of executive orders relating to North Korea, to a person under U.S. jurisdiction is

---

[19] Department of Commerce. Bureau of Industry and Security. 15 C.F.R. Parts 732, 738, 740, 742, 746, 772, and 774. "North Korea: Imposition of New Foreign Policy Controls." 15 C.F.R. Part 746.4(b)(1), for example, provides the following:

...luxury automobiles; yachts; gems; jewelry; other fashion accessories; cosmetics; perfumes; furs; designer clothing; luxury watches; rugs and tapestries; electronic entertainment software and equipment; recreational sports equipment; tobacco; wine and other alcoholic beverages; musical instruments; art; and antiques and collectible items, including but not limited to rare coins and stamps are subject to a general policy of denial. For further information on luxury goods, see supplement no. 1 to part 746.

This language generally is stated at 15 C.F.R. Part 746.4(b)(1). See also Supplement No. 1 to 15 C.F.R. Part 746— Examples of Luxury Goods following 15 C.F.R. Part 746. 15 C.F.R. Part 700—the Export Administration Regulations (EAR)—are likely to be rewritten or substantially amended in the near future to reflect to enactment of the Export Controls Act of 2018; see footnote 17.

[20] 15 C.F.R. Part 746.4(b)(4) provides that licenses "are subject to a general policy of approval" if the intended export is a humanitarian item "(e.g., blankets, basic footware, heating oil, and other items meeting subsistence needs) intended for the benefit of the North Korean people; items in support of United Nations humanitarian efforts; and agricultural commodities or medical devices items that are determined by BIS [Bureau of Industry and Security], in consultation with the interagency license review community, not to be luxury goods are subject to a general policy of approval." 15 C.F.R. Part 746.4(c) provides that some licensing is possible for items used by news media, U.S. government, International Atomic Energy Agency (IAEA), the European Atomic Energy Community (Euratom), safe operation of civil aircraft, operation technology related to other legally exported commodities, and some gift parcels if no luxury goods are included. 15 C.F.R. Part 700—the Export Administration Regulations (EAR)—are likely to be rewritten or substantially amended in the near future to reflect to enactment of the Export Controls Act of 2018; see footnote 17.

prohibited.[21] The President, in Executive Order 13570 of April 18, 2011, stated the following:

> Except to the extent provided in statutes or in licenses, regulations, orders, or directives that may be issued pursuant to this order, and notwithstanding any contract entered into or any license or permit granted prior to the date of this order, the importation into the United States, directly or indirectly, of any goods, services, or technology from North Korea is prohibited.[22]

United States persons are also prohibited from registering a vessel in North Korea; obtaining authorization to fly the North Korean flag on a vessel; or owning, leasing, operating, or insuring any vessel so flagged.[23]

Terrorism

The Government of North Korea is designated as a state sponsor of acts of international terrorism under several statutes governing U.S. export policy, arms trade, and foreign assistance.[24]

---

[21] 31 C.F.R. Part 510, reissued February 22, 2018; 83 F.R. 9182.

[22] Section1, Executive Order 13570, 76 F.R. 22291, April 18, 2011.

[23] Executive Order 13466, "Continuing Certain Restrictions With Respect to North Korea and North Korean Nationals," 73 F.R. 36787, June 26, 2008. See also 31 C.F.R. Part 510, issued November 4, 2010.

[24] Originally designated pursuant to Section 6(j) of the Export Administration Act of 1979 (EAA'79; P.L. 96-72; 50 U.S.C. app. 2405). The Export Controls Act of 2018 (ECA'18; Part I, Subtitle B, Title XVII, of the John S. McCain National Defense Authorization Act for Fiscal Year 2019; P.L. 115-232; 50 U.S.C. 4801 et seq.) repealed all but three sections of the Export Administration Act of 1979, including Section 6(j). In its place, Section 1768, ECA'18 (50 U.S.C. 4826) continues under Section 1754(c) (50 U.S.C. 4813), any designation or related determination, delegation, rule, regulation, license, or order made under Section 6(j), EAA'79. In addition, Section 1768, ECA'18, provides that any reference elsewhere in law to Section 6(j), EAA'79 shall be deemed to be a reference to Section 1754(c), ECA'18. Also currently listed as supporters of international terrorism are Iran, Syria, and Sudan. See also Section 40 of the Arms Export Control Act (22 U.S.C. 2780) and Section 620A of the Foreign Assistance Act of 1961 (22 U.S.C. 2371). Each of these sections of law authorizes the promulgation of a list of supporters of international terrorism, though no list has been generated under either section. It is generally considered that the list maintained pursuant to the Export Administration Act of 1979, and now the Export Controls Act of 2018, applies to all three laws.

Then-Secretary of State Rex Tillerson designated the Government of North Korea as a state sponsor of acts of international terrorism on November 17, 2017: Department of State Public Notice 10211, "Democratic People's Republic of Korea (DPRK) Designation as a State

Most elements of trade, Beneficiary Developing Country status, sales of items on the U.S. Munitions List, most foreign aid, Export-Import Bank funding, and support in international financial institutions are denied to countries found to be supporting international terrorism under the Export Administration Act of 1979 (now superseded by the Export Controls Act of 2018 [ECA'18]). North Korea was added to the list effective January 20, 1988. Placement on the Section 6(j) (now Section 1754(c) of the ECA'18) list not only results in the constriction of trade possibilities; placement also may trigger denial of beneficial trade designation (NTR or GSP), unfavorable tax status for investors, new limits on diplomatic relations, opposition in international financial institutions, and stricter licensing requirements for trade with the United States in food and medicine.[25]

North Korea is also among those countries listed as being in violation of Section 40A of the Arms Export Control Act, which prohibits the selling or licensing of defense articles or defense services to any country that the President finds "is not cooperating fully with United States antiterrorism

---

Sponsor of Terrorism (SST)," November 27, 2017, 82 F.R. 56100. The timing suggests that this step was taken largely in response to a requirement from Congress for the Secretary to determine, not later than early November 2017, "whether North Korea meets the criteria for designation as a state sponsor of terrorism." Section 324, Korean Interdiction and Modernization of Sanctions Act (Title III, Countering America's Adversaries Through Sanctions Act; P.L. 115-44; 131 Stat. 954). Previously, Secretary of State George Shultz had designated the DPRK government on January 20, 1988, following the mid-air bombing of Korean Air Lines flight 858, reportedly brought about by DPRK government agents in late 1987. On June 26, 2008, President George W. Bush certified that the Government of North Korea had met the requirements of U.S. law to be found to no longer support acts of international terrorism. Memorandum of June 26, 2008; 73 F.R. 37351. Secretary of State Condoleeza Rice, a few months later, rescinded North Korea's listing as a terrorist supporter, as required by law. Department of State Public Notice 6415; October 11, 2008; 73 F.R. 63540. CRS Report RL31696, *North Korea: Economic Sanctions Prior to Removal from Terrorism Designation*, and CRS Report R43835, *State Sponsors of Acts of International Terrorism—Legislative Parameters: In Brief*, provide details on the statutes relating to a terrorism designation.

[25] See Section 502 of the Trade Act of 1974 (P.L. 93-618;19 U.S.C. 2462); Section 901(j) of the International Revenue Code (26 U.S.C. 901(j)); 10 United States Code §2327 (General Military Law); Section 40 of the State Department Basic Authorities Act of 1956 (P.L. 84-885; 22 U.S.C. 2712); Section 205 of the State Department Basic Authorities Act of 1956 (P.L. 84-885; 22 U.S.C. 4305); Section 1621 of the International Financial Institutions Act (P.L. 95-118; 22 U.S.C. 262p-4q); Section 501 of Miscellaneous Appropriations, 2000 (H.R. 3425, enacted by reference in P.L. 106- 113; 22 U.S.C. 2395a note); and Section 906 of the Trade Sanctions Reform and Export Enhancement Act (H.R. 5426, enacted by reference in P.L. 106-387; 22 U.S.C. 7205), as amended.

efforts." The President is required to make such a determination annually, and the prohibition may be waived on grounds that it is in the national interest to do so.[26]

Nonmarket Economy

The Trade Agreement Extension Act of 1951 required the suspension of Most-Favored-Nation trade status (MFN, which is now known as Normal Trade Relations [NTR]) for all Communist countries except Yugoslavia. As a result, North Korea was denied MFN trade status on September 1, 1951.

North Korea remains listed in the headnotes of the Harmonized Tariff Schedule of the United States (HTSUS) as a Rate of Duty Column 2 country (along with Cuba). As a result, while trade is not prohibited with North Korea under the relevant trade laws, tariffs are set at the highest rates for imports from that country.[27] A side result of being denied MFN or NTR status is that any such country is also denied preferential trade treatment under the Generalized System of Preferences (GSP), pursuant to the Trade Act of 1974.[28] As a nonmarket economy found to deny its citizens the right or opportunity to emigrate, North Korea is not eligible to participate in any U.S. government program that makes credit, credit guarantees, or investment guarantees available, nor may the President enter into any commercial agreement with the country.[29]

Proliferator

On several occasions, North Korean entities have been found to be in violation of U.S. missile nonproliferation laws.[30] Once a finding is made,

---

[26] 22 U.S.C. 2781. The most recent certification, Department of State Public Notice 10428 of May 5, 2018 (83 F.R. 23988; May 23, 2018), includes Eritrea, Iran, North Korea, Syria, and Venezuela. This section of law was added to the Arms Export Control Act in late 1996; North Korea has been included on the list each year since its inception.

[27] Harmonized Tariff Schedule of the United States, general note 3(b).

[28] Section 502(b)(1) of P.L. 93-618 (19 U.S.C. 2461).

[29] Section 402 of the Trade Act of 1974, popularly referred to as the Jackson-Vanik amendment (19 U.S.C. 2432), and Section 409 of that Act (19 U.S.C. 2439).

[30] Section 73 of the Arms Export Control Act (P.L. 90-629; 22 U.S.C. 2797b), Section 11B of the Export Administration Act (P.L. 96-72; 50 U.S.C. App. 2410b), and Sections 2 and 3 of the Iran, North Korea, and Syria Nonproliferation Act of 2000 (P.L. 106-178; 50

the imposition of sanctions is mandatory, though sanctions may be waived if the President finds it "essential to the national security of the United States" to do so. The severity of the sanction depends on the type of material or technology transferred. The duration of the sanction also depends on the material or technology involved; generally sanctions are imposed for two years or more.

Sanctions include, at a minimum, a denial of contracts with agencies of the U.S. government, denial of licenses for items on the U.S. Munitions List (USML),[31] and, at a maximum, a denial of all licenses for importing into the United States for the foreign person or entity.

Because North Korea is a nonmarket economy,[32] all relevant activities of the Government of North Korea are also sanctioned when entities in North Korea are found to have engaged in proliferation under U.S. law.

With the nuclear weapons test of October 8, 2006, President Bush exercised the authority granted his office to cut off all foreign aid except humanitarian and food aid, deny sales or transfers of defense articles and defense services, deny export licenses for items on the USML, deny foreign military financing, deny credit underwritten or provided by government

---

U.S.C. 1701 note), as amended. North Korea was added to the latter Act on October 13, 2006, with the signing into law of the North Korea Nonproliferation Act of 2006 (P.L. 109-353; 120 Stat. 2015). Section 1766 of the ECA'18 repealed all but three sections of the Export Administration Act of 1979; the remaining three—Sections 11A, 11B, and 11C—are to be continued under the authorities of the National Emergencies Act (P.L. 94-412; 50 U.S.C. 1601) and International Emergency Economic Powers Act (P.L. 95-223; 50 U.S.C. 1702).

31 "In furtherance of world peace and the security and foreign policy of the United States, the President is authorized to control the import and export of defense articles and defense services and to provide foreign policy guidance to persons of the United States involved in the export and import of such articles and services." To accomplish this, the President is authorized to designate items to be controlled—the United States Munitions List (USML). Section 38(a)(1) of the Arms Export Control Act (P.L. 90-629; 22 U.S.C. 2778(a)(1)). The USML may be found at 22 C.F.R. Part 121.1.

32 Section 74(a)(8)(B) of the Arms Export Control Act (P.L. 90-629; 22 U.S.C. 2797c(a)(8)(B)) applies restrictions to a government of a country deemed to be a nonmarket economy when an entity under the jurisdiction of that government engages in missile proliferation activities, because the separation between government and commerce is not distinct. Section 74 provides, in part: "...in the case of countries with non-market economies ... the term 'person' means ... all activities of that government relating to the development or production of any missile equipment or technology; and ... all activities of that government affecting the development or production of electronics, space systems or equipment, and military aircraft...."

coffers, withhold U.S. support in the international financial institutions, deny export licenses for dual-use items, and withhold Export-Import Bank support.[33] At the time, the United States already maintained a fairly comprehensive sanctions regime on North Korea, thus most of these relationships were already broken or limited.

Access to Landing or Ports

President Trump, in Executive Order 13810 of late 2017, determined that "no aircraft in which a foreign person has an interest that has landed" in North Korea would be permitted to land in the United States within 180 days from its departure from North Korea. Seagoing vessels are similarly restricted. The order allows for completion of standing contracts.[34]

Aid

Terrorism

Because the Secretary of State has identified the Government of North Korea as a state sponsor of acts of international terrorism, it is denied most aid under the Foreign Assistance Act of 1961.[35] North Korea's access to U.S. foreign assistance is also limited in annual foreign operations

---

[33] Section 102 of the Arms Export Control Act (P.L. 90-629; 22 U.S.C. 2799aa-1), popularly referred to as the Glenn amendment; Section 2(b)(4) of the Export-Import Bank Act of 1945 (P.L. 79-173; 12 U.S.C. 635(b)(4)); and the Department of State, Foreign Operations, and Related Programs Appropriations Act, 2019 (Division F of the Consolidated Appropriations Act, 2019; P.L. 116-6; 133 Stat. 13), relating to Export-Import Bank funding. On December 7, 2006, President Bush determined that North Korea, a non-nuclear-weapon state, had detonated a nuclear explosive device, citing Section 102(b) of the Arms Export Control Act and Section 129 of the Atomic Energy Act. Presidential Determination No. 2007-07. *Public Papers of the President.* December 18, 2006.

[34] Section 2, Executive Order 13810 of September 20, 2017; 82 F.R. 44706.

[35] Section 620A of that Foreign Assistance Act of 1961 (P.L. 87-195; 22 U.S.C. 2371) prohibits most aid under its auspices as well as that of the Agricultural Trade Development and Assistance Act of 1954 (P.L. 480; 7 U.S.C. 1691 et seq.), Peace Corps Act (P.L. 87-293; 22 U.S.C. 2501 et seq.), and Export-Import Bank Act of 1945 (P.L. 79-173; 12 U.S.C. 635 et seq.). The section, however, also provides the President the authority to waive its application on humanitarian or national security grounds in some instances.

appropriations measures.[36] In most instances, it is not expressly stated that the restriction or prohibition is associated with North Korea's place on the terrorist list. Even if the appropriations law prohibits the availability of foreign aid, there are numerous exceptions to the law. Thus, programs in nonproliferation, demining, child survival, conservation and biodiversity, food aid, debt buybacks, health and disease prevention, unanticipated contingencies, international disaster assistance, and antiterrorism may be funded or supported in spite of country-specific restrictions.

Under the Department of State, Foreign Operations, and Related Programs Appropriations Act,[37] North Korea is generally denied direct foreign aid, economic support funds (ESF) for energy-related programs, and direct loans, credits, insurance and guarantees of the Export-Import Bank.

---

[36] When appropriations law prohibits the availability of foreign aid, however, numerous exceptions to the law allow aid to be made available for targeted programs. Thus, programs in nonproliferation, demining, child survival, conservation and biodiversity, food aid, debt buybacks, health and disease prevention, unanticipated contingencies, international disaster assistance, and antiterrorism may be funded or supported in spite of country-specific restrictions. The President also is authorized, under Section 614 of the Foreign Assistance Act of 1961 (22 U.S.C. 2364) to furnish foreign aid "without regard to any provision of this Act, the Arms Export Control Act, any law relating to receipts and credits accruing to the United States, and any Act authorizing or appropriating funds for use under this Act...."

[37] Division F of the Consolidated Appropriations Act, 2019 (P.L. 116-6; 133 Stat. 13). See particularly Section 7007— Prohibition Against Direct Funding for Certain Countries; Section 7021—Prohibition on Assistance to Governments Supporting International Terrorism; and Section 7043(d)— East Asia and the Pacific, which continues funding for the State Department's database of North Korea's gulags and prisons, begun by Section 7032(i) of P.L. 113-76 (128 Stat. 513), and allows funding for the promotion of human rights in North Korea.

Subsequent to its nuclear tests, North Korea would also be denied U.S. Export-Import Bank support under this Act— see Title VI, paragraph providing for the program account of the Ex-Im Bank. In the North Korean Human Rights Act of 2004 (P.L. 108-333; 22 U.S.C. 7801 et seq.), as amended, Congress authorizes funding for FY2013 through 2017 for various aid and diplomacy programs for refugees from North Korea (migration and refugee assistance); up to $2 million per year for broadcasting into the country; $2 million per year in economic support funds for programs promoting democracy, human rights, and governance; and $5 million in Economic Support Funds to promote human rights, address needs of North Korean refugees, improve accountability of humanitarian assistance inside the country, improve the flow of information into and out of the country, and promote a peaceful reunification of the peninsula under a democratic government. Section 403 of the North Korea Sanctions and Policy Enhancement Act of 2016 (22 U.S.C. 9253) authorized to be appropriated $10 million through FY2021 for various programs stated in the 2004 and 2016 Acts.

The prohibitions on direct foreign aid to North Korea also make that country ineligible for Millennium Challenge Account programs.[38]

At the President's discretion, North Korea is also subject to the economic sanctions provided in three provisions of law addressing human rights conditions: the Foreign Assistance Act of 1961,[39] under which North Korea is annually castigated for its human rights record; the International Religious Freedom Act of 1998,[40] under which the administration has identified North Korea as a "country of particular concern" since 2001; and the Trafficking Victims Protection Act of 2000,[41] under which the administration has, since 2003, classified North Korea as a Tier 3 (most severe) offender of standards pertaining to the trafficking of persons for slavery or sex trade. Because of North Korea's failure to comply with minimum standards relating to trafficking in persons, President Obama, in 2010, strengthened the sanctions against North Korea to deny foreign assistance and also to deny "funding for participation by officials or employees of such governments in educational and cultural exchange programs for the subsequent fiscal year."[42] Any sanctions imposed pursuant to these acts would be largely redundant, however, with penalties already prescribed to North Korea for the above-stated reasons.

Under Department of Defense Appropriations, 2018, North Korea is denied assistance under that act "unless specifically appropriated for that purpose."[43]

---

[38] Millennium Challenge Act of 2003 (Division D of Title VI of P.L. 108-199; 22 U.S.C. 7701 et seq.); Millennium Challenge Corporation, "Report on the Criteria and Methodology for Determining the Eligibility of Candidate Countries for Millennium Challenge Account Assistance in in Fiscal Year 2018," Notice of September 29, 2017; 82 F.R. 46289.

[39] Sections 116 and 502B of P.L. 87-195 (22 U.S.C. 2151n and 2304, respectively), as amended.

[40] P.L. 105-292 (22 U.S.C. 6401 et seq.). Department of State. "Secretary of State's Determination Under the International Religious Freedom Act of 1998," Public Notice 10635; 83 F.R. 65782; December 21, 2018.

[41] P.L. 106-386 (22 U.S.C. 7101 et seq.). Department of State. *Trafficking in Persons Report*. June 2018. https://www.state.gov/j/tip/rls/tiprpt/index.htm.

[42] Section 110(d)(1)(A)(ii) of P.L. 106-386; Presidential Determination No. 2011-15 of September 10, 2010 (75 F.R. 67017).

[43] Section 8043 of the Department of Defense Appropriations, 2019 (Division A of P.L. 115-245; 132 Stat. 3011).

Nonmarket Economy

The Export-Import Bank Act of 1945 singles out Marxist-Leninist countries for denial of guarantees, insurance, credit, or other Bank funding programs. North Korea is specifically cited as a Marxist-Leninist country for purposes of the Export-Import Bank.[44]

The Foreign Assistance Act of 1961 denies most nonhumanitarian foreign assistance to any Communist country. North Korea is among five countries so designated, though the law is not limited to those countries named.[45]

Several laws deny benefits or assistance to Communist countries, but do not explicitly name any particular state. Because North Korea has been denied such benefits or aid in the course of the events of the early 1950s and thereafter, these other sections of law would probably be redundant if applied to or cited for North Korea.

In some instances, the President may determine that, for purposes of a particular law, North Korea is no longer a "Marxist-Leninist state." If, however, all other aspects of the U.S.-North Korea relationship were to improve, it would probably be necessary for Congress to remove North Korea from the list set out in the Export-Import Bank Act and the Foreign Assistance Act of 1961, or necessary for the President to exercise waiver authority made available to his office under those acts, to make these other laws inapplicable to North Korea.[46]

---

[44] Section 2(b)(2) of P.L. 79-173 (12 U.S.C. 635(b)(2)); amended in 1986 to include this ban on funding to Marxist-Leninist states.

[45] Section 620(f) of P.L. 87-195 (22 U.S.C. 2370(f)). Consider also subsec. (h) of that section, which requires the President to "adopt regulations and establish procedures to insure that United States foreign aid is not used in a manner which, contrary to the best interests of the United States, promotes or assists the foreign aid projects or activities of any country that is a Communist country for purposes of subsection (f)."

[46] For example: Section 620(h) of the Foreign Assistance Act of 1961 (P.L. 87-195; 22 U.S.C. 2370(h)), Sections 502(b)(1) and (b)(2)(A) of the Trade Act of 1974 (P.L. 93-618; 19 U.S.C. 2462(b)(2)(A)), and Section 43 of the Bretton Woods Agreements Act (P.L. 79-171; 22 U.S.C. 286aa), the latter of which requires the U.S. Executive Directors to the International Monetary Fund "to actively oppose any facility involving use of Fund credit by any Communist dictatorship...."

Arms Sales and Arms Transfers

The International Traffic in Arms Regulations (ITAR), administered by the Department of State, begins "It is the policy of the United States to deny licenses and other approvals for exports and imports of defense articles and defense services, destined for or originating in certain countries." Reasons include requirements of the U.N. Security Council, terrorism, and policies that include arms embargoes and sanctions. Countries broadly restricted for arms trade include Belarus, Burma, China, Cuba, Iran, North Korea, Syria, and Venezuela.[47]

The first ITAR was issued on August 26, 1955; North Korea has been listed as a restricted country from the ITAR's inception. North Korea is also restricted under ITAR as the United States denies North Korea conventional arms to comply with U.N. Security Council requirements.[48]

Importing of defense articles and defense services is similarly restricted by the Department of Justice's Bureau of Alcohol, Tobacco, Firearms, and Explosives, the regulations of which state the following:

> It is the policy of the United States to deny licenses and other approvals with respect to defense articles and defense services originating in certain countries or areas. This policy applies to Afghanistan, Belarus (one of the states composing the former Soviet Union), Cuba, Iran, Iraq, Libya, Mongolia, North Korea, Sudan, Syria, and Vietnam. This policy applies to countries or areas with respect to which the United States maintains an arms embargo (e.g., Burma, China, the Democratic Republic of the Congo, Haiti, Liberia, Rwanda, Somalia, Sudan, and UNITA (Angola)). It also applies when an import would not be in furtherance of world peace and the security and foreign policy of the United States.[49]

---

[47] 22 C.F.R. Part 126.1(a) and (d), authorized pursuant to Section 38 of the Arms Export Control Act (P.L. 90-629; 22 U.S.C. 2778).

[48] U.N. Security Council Resolution 1718 (2006), October 14, 2006 (U.N. document S/Res/1718 (2006)); U.N. Security Council Resolution 1874 (2009), June 12, 2009 (U.N. document S/Res/1874 (2009)).

[49] 27 C.F.R. Part 447.52(a), also authorized under Section 38 of the Arms Export Control Act; and President Barack Obama, "Administration of Reformed Export Controls," Executive Order 13637 of March 8, 2013, 78 F.R. 16129.

Again, the President has the authority to change these regulations by removing North Korea from the list of restricted countries.

Access to Assets

Declaration of National Emergency[50]

On June 26, 2008, when the Six Party Talks appeared to be making progress, President Bush determined it was no longer in the national interest to continue certain restrictions imposed on trade and transactions with North Korea, in place since 1950.[51] At the same time, however, he found that[52]

> ... the current existence and risk of the proliferation of weapons-usable fissile material on the Korean Peninsula constitute an unusual and extraordinary threat to the national security and foreign policy of the United States, and I hereby declare a national emergency to deal with that threat. I further find that, as we deal with that threat through multilateral diplomacy, it is necessary to continue certain restrictions with respect to

---

[50] For background on the use of national emergency and international economic emergency authorities, see CRS Report R45618, *The International Emergency Economic Powers Act: Origins, Evolution, and Use*, coordinated by Christopher A. Casey.

[51] The President. "Termination of the Exercise of Authorities Under the Trading With the Enemy Act With Respect to North Korea," Proclamation 8271 (June 26, 2008; 73 F.R. 36785). On December 16, 1950, President Truman invoked authority granted his office under the Trading With the Enemy Act (TWEA) to declare that a U.S. national emergency existed because of the outbreak of the Korean War (and events elsewhere, as "world conquest by communist imperialism is a goal of the forces of aggression that have been loosed upon the world") (Proclamation 2914; 15 F.R. 9029). A few days later, the Department of the Treasury issued Foreign Assets Control Regulations (FACR; 31 C.F.R. Part 500; 15 F.R. 9040, December 19, 1950, and subsequently amended) to forbid any financial transactions involving, or on behalf of, North Korea and China, including transactions related to travel or the access to North Korean assets that were subject to U.S. jurisdiction. Korea-related FACR have been modified on numerous occasions to take into consideration new circumstances (i.e., transactions relating to technology not in existence at the time the regulations were issued) or to ease restrictions in response to changing conditions (i.e., signing of the Agreed Framework, emerging reports of famine, North Korea's announced moratorium on missile testing).

[52] Executive Order 13466, "Continuing Certain Restrictions With Respect to North Korea and North Korean Nationals," 73 F.R. 36787, June 26, 2008. 31 C.F.R. Part 510, November 4, 2010.

North Korea that would otherwise be lifted pursuant to a forthcoming proclamation that will terminate the exercise of authorities under the Trading With the Enemy Act.... Accordingly, I hereby order... the following are blocked and may not be transferred, paid, exported, withdrawn, or otherwise dealt in:

- all property and interests in property of North Korea or a North Korean national that ... were blocked as of June 16, 2000,[53] and remained blocked immediately prior to the date of this order.
- United States persons may not register a vessel in North Korea, obtain authorization for a vessel to fly the North Korean flag, or own, lease, operate, or insure any vessel flagged by North Korea.

In a series of executive orders, the Presidents in succession have used the national emergency to block access to assets of designated individuals and entities

- Executive Order 13466 of June 26, 2008 (President Bush; 73 F.R. 36787)— continuing the block of assets that were blocked as of June 16, 2000;
- Executive Order 13551 of August 30, 2010 (President Obama; 75 F.R. 53837)— blocking assets of Kim Yong Chol, and the entities Green Pine Associated Corporation, Reconnaissance General Bureau, and Office 39;[54]
- Executive Order 13570 of April 18, 2011 (President Obama; 76 F.R. 22291)—to prohibit the import of goods, services, or technology form North Korea;
- Executive Order 13687 of January 2, 2015 (President Obama; 81 F.R. 14943)— blocking assets that come under U.S. jurisdiction of

[53] In 1999, President Clinton announced he would lift many restrictions on U.S. exports to and imports from North Korea in areas other than those controlled for national security concerns; the Departments of Commerce, Treasury, and Transportation issued new regulations a year later, effective June 16, 2000, to implement the new policy.

[54] OFAC may add to this list at any time. The Specially Designated Nationals list may be searched online at http://www.treasury.gov/resource-center/sanctions/SDN-List/Pages/default.aspx.

any agency, instrumentality, or controlled entity of the North Korean government or the Workers' Party of Korea, or any official of those entities; those found to materially support transactions with or act on behalf of those blocked entities;[55]

- Executive Order 13722 of March 15, 2016 (President Obama; 81 F.R. 14943)— blocking property and interests in property that come under U.S. jurisdiction of the Government of North Korea, the Workers' Party of Korea, and designated entities that operate in DPRK's transportation, mining, energy, or financial services sectors (or other sectors as designated). Designees include Kim Jongun;[56] and
- Executive Order 13810 of September 20, 2017 (President Trump; 82 F.R. 44705)—blocking property and interests in property that come under U.S. jurisdiction of designated persons and entities that operate in the DPRK's "construction, energy, financial services, fishing, information technology, manufacturing, medical, mining, textiles, or transportation industries"; persons who own, control, or operate any of North Korea's sea-, land-, or airports; or those who generate income for the government or Workers' Party.

Generally, the President has the authority to change regulations, as long as those changes meet the requirements of any relevant law. He must also annually revisit his declaration of a state of national emergency; it

---

[55] OFAC, on January 2, 2015, designated three entities and 10 individuals under this executive order: Reconnaissance General Bureau, Korea Mining Development Trading Corporation, and Korea Tangun Trading Corporation; and Kil Jong Hun, Kim Kwang Yon, Jang Song Chol, Yu Kwang Ho, Kim Yong Chol, Jang Yong Son, Kim Kyu, Ryu Jin, Kang Ryong, Kim Kwang Chun—all officers of either the government or the sanctioned entities. Department of the Treasury. "Treasury Imposes Sanctions Against the Government of the Democratic People's Republic of Korea," press release, January 2, 2015, Department of the Treasury Notice, 80 F.R. 13667, March 16, 2015. Treasury designated an additional four individuals—Sok Chol Kim, Kwang Hyok Kim, Chong Chol Ri, and Su Man Hwang—and one entity— EKO Development and Investment Company—effective November 13, 2015. Department of the Treasury Notice, November 18, 2015, 80 F.R. 72147.

[56] Pursuant to Executive Order 13722, OFAC has designated 114 entities, vessels, aircraft, and individuals, including Kim Jong-un, chairman of the Korea Workers' Party and president of the country, national banks, shipping and trading companies, the Ministry of State Security, the Ministry of People's Security Correctional Bureau, and others.

expires if the President does not renew it. He could allow the declaration to expire, or he could lift it at any time. And Congress could terminate a declaration of national emergency by passing a joint resolution under terms of the National Emergencies Act.

Proliferation of Weapons of Mass Destruction

On June 28, 2005, President George W. Bush expanded the authority granted his office to address the threat posed by the proliferation of weapons of mass destruction—authority first exercised by President George H. W. Bush in 1990—to freeze assets and property of those engaged in the proliferation of weapons of mass destruction. The 41st President had declared that the United States faced a national emergency relating to weapons proliferation (in the absence, at the time, of a reauthorized Export Administration Act), and thus took steps in 1990 to control the exports of certain goods and services, and authorized a ban on foreign aid and credit, procurement contracts, imports and exports, support in international financial institutions, and landing rights.[57] Fifteen years later, the 43rd President's executive order took additional steps to block property and assets under U.S. jurisdiction of any person found, in part

> ... to have engaged, or attempted to engage, in activities or transactions that have materially contributed to, or pose a risk of materially contributing to, the proliferation of weapons of mass destruction or their means of delivery (including missiles capable of delivering such weapons), including any efforts to manufacture, acquire, possess, develop, transport, transfer or use such items, by any person or foreign country of proliferation concern;

---

[57] Executive Order 12735 of November 16, 1990. On the same day, the President announced his intention to pocket veto H.R. 4653—the Omnibus Export Amendments Act of 1990—with which Congress intended to reauthorize the then-expired Export Administration Act of 1979. In his announcement, President Bush stated his intentions to curtail significantly trade in goods and services that lent themselves to the proliferation of weapons of mass destruction. See "Memorandum of Disapproval for the Omnibus Export Amendments Act of 1990," *Public Papers of the President*, November 16, 1990. 26 Weekly Comp. Pres. Doc. 1839. E.O. 12735 was subsequently overhauled by President Clinton with the issuance of Executive Order 12938 (November 14, 1994; 59 F.R. 59099; 50 U.S.C. 1701 note). The national emergency therein is renewed annually.

> ... to have provided, or attempted to provide, financial, material, technological or other support for, or goods or services in support of, any activity or transaction described [above] ... or any person whose property and interests in property are blocked pursuant to this order ...[58]

At its outset, Executive Order 13382 identified eight foreign entities as contributors to proliferation, of which three were North Korean. The Office of Foreign Assets Control subsequently added North Korean entities and individuals to this restricted list; of the 442 entities and individuals designated as weapons proliferators, 40 are cited as located in DPRK. In the same findings, OFAC identified entities operating out of Switzerland, Iran, China, South Korea, Russia, Sudan, Syria, United Arab Emirates, Venezuela, Belgium, and Belarus as entities engaging in WMD proliferation.[59]

Counterfeiting and Money-Laundering Banco Delta Asia

On September 12, 2005, the Department of the Treasury found that Banco Delta Asia—a Macau-based bank in which North Korea had holdings of more than $ U.S. 50 million—was a "financial institution of primary money laundering concern."[60] The Department of the Treasury's

---

[58] Executive Order 13382, *Blocking Property of Weapons of Mass Destruction Proliferators and Their Supporters* (June 28, 2005; 70 F.R. 38567).

[59] See Department of the Treasury. Office of Foreign Assets Control. "Nonproliferation: What You Need to Know About Treasury Restrictions." *Information Bulletin*, most recently updated September 19, 2012, http://www.treasury.gov/resource-center/sanctions/Programs/Documents/wmd.pdf. See also testimony of Robert W. Werner, Director of the Office of Foreign Assets Control, Department of the Treasury, before the House Committee on Financial Services, February 16, 2006.

On January 9, 2007, the Department of the Treasury announced that it had designated Bank Sepah, a state-owned Iranian financial institution, as an entity materially contributing to Iran's proliferation activities. Bank Sepah, coincidentally, according to Treasury, is credited with transferring more than $500,000 for an associate of the Korean Mining Development Corporation. That entity was cited on December 28, 2006, under the Iran, North Korea, and Syria Nonproliferation Act of 2000, for exporting missile technology to Iran. Department of the Treasury. Press release. *Iran's Bank Sepah Designated by Treasury: Sepah Facilitating Iran's Weapons Program*. January 9, 2007. HP-219; Weisman, Steven. "U.S. Prohibits All Transactions with a Major Iranian Bank," *The New York Times*. January 10, 2007. p. 3. Korean Mining Development Corporation is cited multiple times under Executive Order 13882.

[60] Pursuant to 31 U.S.C. 5318A, as enacted by the USA PATRIOT Act (Section 311 of P.L. 107-56; 115 Stat. 298). Effective September 12, 2005, the Department of the Treasury's

Financial Crimes Enforcement Center (FinCEN) found that North Korea may reap as much as $500 million annually from counterfeiting, and another $100 million to $200 million annually from narcotics trafficking. The finding authorizes the Secretary of the Treasury to require "special measures"[61] on the part of U.S. financial institutions and financial agencies that involve increased record keeping and reporting on Banco Delta Asia's transactions. Treasury issued a final rule, effective April 18, 2007, to impose the most stringent fifth special measure—to prohibit certain bank transactions—and issued regulations to implement the rule.[62] The finding and initial proposed rulemaking had a chilling effect on Banco Delta Asia's international business relations. Department of the Treasury officials testified that

> some two dozen financial institutions across the globe have voluntarily cut back or terminated their business with North Korea, notably including institutions in China, Japan, Vietnam, Mongolia, and Singapore. The result of these voluntary actions is that it is becoming very difficult for the Kim Jong-Il regime to benefit from its criminal conduct.[63]

North Korea's funds held in Banco Delta Asia were released in 2007.

Financial Crimes Enforcement Network issued a finding (70 F.R. 55214) and a notice of proposed rulemaking (to amend 31 C.F.R. Part 103; 70 F.R. 55217). Documentation on the use of this authority may be found at https://www.treasury.gov/resource-center/terrorist-illicitfinance/311-Actions/Pages/311-Actions.aspx.

[61] 31 U.S.C. 5318A(b) defines "special measures" as (1) record keeping and reporting of certain financial transactions; (2) collection of information relating to beneficial ownership; (3) collection of information relating to certain payable-through accounts; (4) collection of information relating to certain correspondent accounts; and (5) prohibitions or conditions on opening or maintaining in the United States correspondent accounts or payable-through accounts.

[62] U.S. Department of the Treasury. "Financial Crimes Enforcement Network; Amendment to the Bank Secrecy Act Regulations—Imposition of Special Measure Against Banco Delta Asia, Including Its Subsidiaries Delta Asia Credit Limited and Delta Asia Insurance Limited, as a Financial Institution of Primary Money Laundering Concern," Final Rule, 31 C.F.R. Part 103 (72 F.R. 12730).

[63] Senate Committee on Banking, Housing and Urban Affairs hearings, September 12, 2006, testimony of Treasury Deputy Under Secretary Daniel Glaser. *Congressional Quarterly*.

Access to the U.S. Financial System

Money Laundering Concern Determined in 2016

Congress called on the Secretary of the Treasury, in consultation with the Secretary of State and Attorney General, to determine not later than mid-August 2016 "whether reasonable grounds exist for concluding that North Korea is a jurisdiction of primary money laundering concern."[64] On May 27, 2016, Treasury's Acting Director of FinCEN found that such reasonable grounds exist. In December 2016, FinCEN finalized a rule that prohibits U.S. banks from providing U.S. correspondent accounts to third-country banks to process transactions for North Korean financial institutions.[65] The Department of the Treasury, on announcing the finding, noted the following:

> While current U.S. law already generally prohibits U.S. financial institutions from engaging in both direct and indirect transactions with North Korean financial institutions, this NPRM [notice of proposed rulemaking], if finalized, would require U.S. financial institutions to implement additional due diligence measures in order to prevent North Korean banking institutions from gaining improper indirect access to U.S. correspondent accounts. While North Korea's financial institutions do not maintain correspondent accounts with U.S. financial institutions, North Korean financial institutions frequently conduct transactions on behalf of the North Korean government and state-controlled corporations. The NPRM, if finalized, would prohibit the use of third-country banks' U.S. correspondent accounts to process transactions for North Korean financial institutions.[66]

---

[64] Section 201(c), North Korea Sanctions and Policy Enhancement Act of 2016 (P.L. 114-122; 22 U.S.C. 9221(c)).

[65] U.S. Department of the Treasury, Financial Crimes Enforcement Network. "Finding That the Democratic People's Republic of Korea Is a Jurisdiction of Primary Money Laundering Concern," Notice of finding, 81 F.R. 35441; June 2, 2016. U.S. Department of the Treasury, Financial Crimes Enforcement Network. "Imposition of a Special Measure Against North Korea as a Jurisdiction of Primary Money Laundering Concern," Notice of proposed rulemaking, 81 F.R. 35665; June 3, 2016. Final rule entered into effect December 9, 2016; 81 F.R. 78715.

[66] U.S. Department of the Treasury. "Treasury Takes Action to Further Restrict North Korea's Access to the U.S. Financial System," June 1, 2016, press release.

Other Actions Related to Money Laundering

President Obama also addresses money laundering and counterfeiting in Executive Order 13551 as one of North Korea's many objectionable behaviors to be deterred. The order requires the access to property and interests in property be blocked for any individual or entity identified by the Secretary of the Treasury to have, directly or indirectly[67]

- imported, exported, or reexported to, into, or from North Korea any arms or related materiel;
- provided training, advice, or other services or assistance, or engaged in financial transactions, related to the manufacture, maintenance, or use of any arms or related materiel to be imported, exported, or reexported to, into, or from North Korea, or following their importation, exportation, or reexportation to, into, or from North Korea;
- imported, exported, or reexported luxury goods to or into North Korea;
- engaged in money laundering, the counterfeiting of goods or currency, bulk cash smuggling, narcotics trafficking, or other illicit economic activity that involves or supports the Government of North Korea or any senior official thereof;
- materially assisted, sponsored, or provided financial, material, or technological support for, or goods or services to or in support of, prohibited activities or any person whose property and interests in property are blocked; or
- be owned or controlled by, or to have acted or purported to act for or on behalf of, any person whose property and interests in property are blocked.

[67] Executive Order 13551, "Blocking Property of Certain Persons With Respect to North Korea," 75 F.R. 53837, September 1, 2010.

Targeting Foreign Financial Institutions

In September 2017, President Trump authorized the Secretary of the Treasury to "prohibit the opening and impose strict conditions on the maintenance of correspondent accounts or payable-through accounts in the United States" of any foreign financial institution found to have "conducted or facilitated any significant transaction on behalf of any person" already designated for U.S. sanctions, or to have conducted or facilitated such a transaction related to trade with North Korea. In addition, the U.S.-based assets of any foreign financial institution found to engage in such transactions are to be blocked.

Sanctions Required by the United Nations

U.N. Security Council Resolutions—10 have been adopted between 2006 and September 2017— that form the requirements of the U.N. member states often incrementally clarify, expand on, enhance, or strengthen the previous U.N. actions. Resolutions, however, are not revoked until the problem at hand is resolved, nor are they amended (with a few exceptions).

U.N. Security Council resolutions also establish considerable opportunity for member states to appeal to the Sanctions Committee—established to implement the DPRK resolutions with representatives from each of the 15 Security Council states—for an exception prior to executing a transaction. Overall, the Security Council has sought to protect DPRK's civilian population from the consequences of economic sanctions. UNSCR 2371 (2017), agreed to in August 2017, for example, provides the following:

> [The United Nations Security Council] Reaffirms that the measures imposed by resolutions 1718 (2006), 1874 (2009), 2087 (2013), 2094 (2013), 2270 (2016), 2321 (2016), 2356 (2017), and this resolution are not intended to have adverse humanitarian consequences for the civilian population of the DPRK or to affect negatively or restrict those activities,

> including economic activities and cooperation, food aid and humanitarian assistance, that are not prohibited by resolutions..., and the work of international and nongovernmental organizations carrying out assistance and relief activities in the DPRK for the benefit of the civilian population of the DPRK and decides that the Committee may, on a case-by-case basis, exempt any activity from the measures imposed by these resolutions if the committee determines that such an exemption is necessary to facilitate the work of such organizations in the DPRK or for any other purpose consistent with the objectives of these resolutions, and further decides that the measures specified in paragraph 8 (d) of resolution 1718 (2006) shall not apply with respect to financial transactions with the DPRK Foreign Trade Bank or the Korea National Insurance Corporation if such transactions are solely for the operation of diplomatic or consular missions in the DPRK or humanitarian assistance activities that are undertaken by, or in coordination with, the United Nations ....

Similar language to exempt humanitarian activities is provided in each resolution, and such exemptions are stated throughout each resolution but tailored to particular kinds of transactions.

The Sanctions Committee of the Security Council may designate entities, individuals, vessels, and aircraft to be subject to economic sanctions at any time. In addition, beginning in 2013, the Security Council added annexes to resolutions to name designees, as well as to make the ban on exporting luxury goods to DPRK more explicit, and to define goods and services with military applications that are also subject to an export prohibition.

Currently, the United Nations requires its member states to restrict trade and engagement with North Korea as follows:

The U.N. Sanctions Committee, originally established by UNSC Resolution 1718 (2006) and comprising representatives of each of the 15 states serving on the Security Council, as of mid-2018, has designated 75 North Korean entities, 80 individuals, and 88 vessels for sanctions that target their foreign-based assets, freedom to enter into trade and contracts, and ability to travel, as required in the various resolutions. Designees appear to span a range of activities, including trade and development related to missiles, satellites, aerospace, munitions, energy, mining, space technology, shipping, and electronics. Pursuant to Resolution 1874 (2009), the U.N. Secretary-General established a Panel of Experts, with a maximum of seven experts, to analyze reports and make recommendations regarding implementation of UNSC Resolutions against North Korea.[68]

Table 1. U.N. Security Council Sanctions Requirements for North Korea Activities Based on S/Res/1718 (2006), 1874 (2009), 2087 (2012), 2094 (2013), 2270 (2016), 2321 (2016), 2356 (2017), 2371 (2017), 2375 (2017), and 2397 (2017)

Targeted Activities	Measures To Be Taken By U.N. Member States
Arms and related materiel	Prevent the direct or indirect supply, sale, or transfer to the DPRK, through their territories or by their nationals, or using their flag vessels or aircraft, and whether or not originating in their territories, of all arms and related materiel, including small arms and light weapons and their related materiel, a ban on related financial transactions, technical training including hosting of trainers, advisors, or other officials for the purpose of military-, paramilitary-, or police related training, services or assistance related to manufacture, maintenance or use, and with respect to the shipment of items to or from the DPRK for repair, servicing, refurbishing, testing, reverse-engineering and marketing. Prevent the supply, sale, or transfer of any item, even if not covered by the arms embargo, if such an item could directly contribute to the development of the DPRK's operational capabilities of its armed forces, or to exports that support/enhance the capabilities of armed forces of another Member State outside the DPRK. *Exemptions*: "[A]ctivity is exclusively for humanitarian purposes or exclusively for livelihood purposes which will not be used by DPRK individuals or entities to generate revenue, and also not related to any activity prohibited" by any of the relevant UNSC resolutions; or the "[Sanctions] Committee has determined on a case-by-case basis that a particular supply, sale, or transfer would not be contrary to the objectives of" any of the relevant UNSC resolutions.

[68] Full texts of the panel of experts reports can be found at https://www.un.org/sc/suborg/en/sanctions/1718/annualreports. To date, the panel has filed 13 annual reports; the most recent filing was on December 31, 2018.

Table 1. (Continued)

Targeted Activities	Measures To Be Taken By U.N. Member States
Arms and related materiel related to proliferation	Prevent the direct or indirect supply, sale, or transfer of items relevant to nuclear, ballistic missiles, and other weapons of mass destruction-related programs. List of restricted items: https://www.un.org/sc/suborg/en/node/6787 Prevent trade in any item that a Member State determines could contribute to the DPRK's nuclear or ballistic missile programs, other weapons of mass destruction programs, or other prohibited activities. *No exemptions*.
Proliferation networks	Expel DPRK diplomats, government representatives, other DPRK nationals acting in a governmental or representative office capacity, and foreign nationals that are working on behalf or at the direction of a designated person and/or entity or of a person and/or entity assisting in sanctions evasions or violating the resolutions. *Exemptions*: Presence is required to fulfill U.N. business; a judicial process; medical, safety, or other humanitarian purposes; or case-by-case determination "that the expulsion of the individual would be contrary to the objectives of" the relevant UNSC resolutions.
Diplomatic services	Close representative offices of designated persons and entities, as well as on any persons or entities acting on behalf of such designated persons or entities, as well as prohibit them from participating in joint ventures and any other business arrangements. Reduce the number of staff at DPRK diplomatic missions and consular posts and restrict the entry into or transit through territory of DPRK government members, officials, DPRK armed forces, or members/officials which are associated withprohibited programmes or activities. Limit the number of bank accounts to one per DPRK diplomatic mission and consular post, and one per accredited DPRK diplomat and consular officer. Prohibit the DPRK from using real property (owned or leased) in their territory for non-diplomatic or -consular activities purposes.
Transportation, flagged vessels, ports, aircraft	Inspect cargo destined to or originating from the DPRK or brokered by the DPRK that is within or transiting their territories, including cargo on DPRK flagged aircraft or vessels, carried by rail or road, as well as personal luggage. Prohibit leasing or chartering flagged vessels, aircraft or providing crew services to the DPRK, designated persons and entities, or any persons or entities whom the Member State determines have assisted in sanctions evasions or in violation of the resolutions. Prohibit procuring vessel and aircraft services from the DPRK. Deregister any vessel that is owned, operated or crewed by the DPRK; refuse to register any such vessel that is deregistered by another Member State. Deregister any vessels for which there are reasonable grounds to believe are involved in prohibited activities or transporting prohibited items..

Targeted Activities	Measures To Be Taken By U.N. Member States
	Do not register any vessel that has been deregistered by another Member State. Do not provide insurance or reinsurance for such vessels. Prohibit one's nationals, entities, and persons from registering vessels in the DPRK or to obtain authorization for a vessel to use the DPRK flag. Prohibit any DPRK-flagged vessel from owning, leasing, operating, or providing vessel classification, certification, associated service, and insurance under a Member State's authority. Deny permission to any aircraft to take off from, land in or overfly one's territory if there are reasonable grounds to believe that the aircraft contained prohibited items. Deny port entry if there are reasonable grounds that a vessel is owned, controlled, directly or indirectly, by a designated individual and/or entity. Inspect vessels on the high seas if there are reasonable grounds to believe its cargo is prohibited. All States are required to cooperate with inspections. Seize, inspect, and impound any vessel for which there are reasonable grounds to believe its cargo is prohibited. Prohibit ship-to-ship transfers of any goods supplied, sold, or transferred to or from the DPRK. Sanctions Committee may designate specific vessels for restrictions. Exemptions: In limited instances, case-by-case basis, it is determined that "information demonstrating that such activities are exclusively for livelihood purposes which will not be used by DPRK individuals or entities to generate revenue" and sensitive information is removed. Further exemptions: In limited instances, emergency purposes
Bunkering services	Prohibit providing fuel, supplies, other bunkering servicing to DPRK vessels if reasonable grounds and information exist that they are carrying prohibited items. Exemptions: Humanitarian purposes, or inspection is completed and it is determined cargo constitutes "legal economic activities."
Travel	Prevent entry into or transit through one's territories of designated individuals; individuals acting on behalf of or at the direction of designated individuals; any individual whom a State determines is assisting in the evasion of sanctions, violating the provisions of the resolutions, working on behalf/at the direction of designated individuals; and individuals traveling for the purposes of carrying out activities related to the shipment of items for repair, servicing, refurbishing, testing, reverse-engineering, and marketing. Exemptions: Travel is for humanitarian need, to meet religious obligations, or would further the objectives of relevant UNSC resolutions; or presence is required to fulfill a judicial process; medical, safety, or fulfil U.N. commitments.

Table 1. (Continued)

Targeted Activities	Measures To Be Taken By U.N. Member States
Assets	Freeze assets, funds, and economic resources of entities of the Government of the DPRK and Korean Workers' Party, that a Member State determines are associated with prohibited activities, including designated persons and entities, as well as any persons or entities acting on behalf of or at their direction, or those owned or controlled by them. "Assets include tangible, intangible, movable, immovable, actual or potential, which may be used to obtain funds, goods or services, such as vessels, including maritime vessels." Sanctions Committee may designate vessels to be frozen. Vessels controlled or operated by Offshore Marine Management (OMM) are explicitly subject to freeze. Exemptions: To meet basic expenses (food, rent, medicine, taxes, insurance, utilities, legal and banking fees); extraordinary expenses as identified by the Sanctions Committee; and satisfying a lien judgment; assets applied to DPRK's participation in U.N.; or case-by-case determinations relating to the costs of delivering humanitarian assistance, or denuclearization.
Other finances	Prevent providing financial services, including bulk cash and gold, the opening of banking subsidiaries, public financial support, new commitments for grants, and financial assistance or concessional loans that could contribute to the DPRK's prohibited programs/activities, or to the evasion of sanctions. Prohibit opening of new branches, subsidiaries, and representative offices of DPRK banks; close existing branches, subsidiaries and representative offices; and terminate joint ventures, ownership interests or correspondent banking relationships with DPRK banks. Prohibit opening new representative offices, subsidiaries, or bank accounts in the DPRK. Close existing offices, subsidiaries and banking accounts in the DPRK if there are reasonable grounds that it could contribute to DPRK's prohibited programs. Prohibit public and private financial support for trade with the DPRK, including granting of export credits, guarantees, or insurance to one's nationals, or entities involved in such trade. Prohibit the opening, maintenance, and operation of all joint ventures or cooperative entities, new or existing, with DPRK entities or individuals. Close any existing joint venture or cooperative entity within 120 days of September 11, 2017, unless approved by the committee on a case-by-case basis. Exemptions: Humanitarian and developmental purposes "directly addressing the needs of the civilian population, or the promotion of denuclearization"; activities under the Vienna Convention on Diplomatic Relations; or wind-down costs of closing branches. Further exemptions: Does not apply to existing China-DPRK hydroelectric power infrastructure projects or the Russia-DPRK Rajin-Khasan port and rail project, intended solely to transport Russia-origin coal.

Targeted Activities	Measures To Be Taken By U.N. Member States
Specialized teaching and training; scientific and technical cooperation	Prevent specialized teaching or training of DPRK nationals of disciplines which could contribute to the DPRK's proliferation-sensitive nuclear activities and development of nuclear weapon delivery systems, including teaching of advanced physics, advanced computer simulation and related computer sciences, geospatial navigation, nuclear engineering, aerospace engineering, aeronautical engineering, and related disciplines. Suspend scientific and technical cooperation involving persons or groups officially sponsored by or representing the DPRK except for medical exchanges. Exemptions: Determined on a case-by-case basis by the committee (fields of nuclear science and technology, aerospace and aeronautical engineering and technology, advanced manufacturing production techniques and methods); and determined by the engaging State and notified to the committee in advance of such determination (all other scientific or technical cooperation).
Coal, minerals, seafood, textiles, statues, new helicopters, and other sectors ban	Prohibit the DPRK from supplying, selling, transferring, directly or indirectly, of coal, iron and iron ore, gold, titanium ore, vanadium ore, copper, nickel, silver, zinc, and rare earth minerals, lead and lead ore, food and agricultural products, machinery, electrical equipment, earth and stone including magnesite and magnesia, wood and vessels. Prohibit procuring such material from the DPRK, by their nationals, or by using their flag vessels or aircraft, whether or not originating in the territory of the DPRK. Prohibit the supply, sale, or transfer to the DPRK of all industrial machinery, transportation vehicles, iron, steel, and other metals. Prohibit export of seafood from DPRK. Prohibit export of textiles from DPRK. Prohibit the supply, sale, or transfer of DPRK statuary. Prohibit the supply, sale, or transfer to DPRK of new helicopters, and new or used vessels. Exemptions: Only applicable to iron and iron ore: contracts had 30 days to be completed (from August 5, 2017); coal transit through the port of Rajin (Rason) requires prenotification to Sanctions Committee. Further exemption: Spare parts needed to maintain the safe operation of DPRK commercial civilian passenger aircraft. Case-by-case determinations by the Sanctions Committee for trade in vessels. Exemption for seafood: Case-by-case approval by Sanctions Committee of seafood contracts completion within 45 days of August 5, 2017. Exemption for textiles: Case-by-case approval by Sanctions Committee of textile contracts completion within 135 days of September 11, 2017.
Fuel ban	Prohibit selling or supplying aviation fuel, jet fuel, rocket fuel, condensates, and natural gas liquids to the DPRK. Prohibit the supply, sale, or transfer to the DPRK of all refined petroleum products in excess of the aggregate amount of 500,000 barrels during periods of 12 months beginning on January 1, 2018.

Table 1. (Continued)

Targeted Activities	Measures To Be Taken By U.N. Member States
	Prohibit the supply, sale, or transfer of crude oil that exceeds the aggregate amounts of 4 million barrels or 525,000 tons per 12-month periods from December 22, 2017. Member States are required to report the amount of crude oil provided to the DPRK to the 1718 Committee every 90 days. Exemptions: Case-by-case basis determination of "verified essential humanitarian needs"; and civilian passenger aircraft outside DPRK for round-trip flights to/from DPRK. Exemption for crude oil: Case-by-case basis determination of crude oil shipments "which is exclusively for livelihood purposes of DPRK nationals." Exemption for refined petroleum products: Case-by-case basis determination of refined petroleum products shipments are "exclusively for livelihood purposes of DPRK nationals."
Exported labor	Prohibit providing work authorizations for DPRK nationals. Repatriate to the DPRK all DPRK nationals earning income in their jurisdiction and all DPRK government safety oversight attachés within 24 months from December 22, 2017. Exemption: Case-by-case determinations for delivery of humanitarian assistance, denuclearization, or other services related to the objectives of UNSC resolutions. Further exemption: Does not apply to contracts already finalized. Exemption relating to repatriation: Member state may take into account international law, including laws relating to refugees and international human rights, when considering repatriation.
Targeted Activities	Measures To Be Taken By U.N. Member States
Luxury goods ban	Prevent direct or indirect supply, sale, or transfer to the DPRK of luxury goods, including those items listed in Annex IV of resolution 2094 (2013), Annex IV of resolution 2270 (2016), and Annex IV of resolution 2321 (2016). No exemptions.

Source: U.N. Security Council 1718 Committee: https://www.un.org/sc/suborg/en/sanctions/1718. CRS edited for length.

CONCLUSION

The U.S. economic sanctions imposed on North Korea exemplify both the independent and intertwined aspects of the relationship between the legislative and executive branches. Congress defers the broadest power to the President, in the National Emergencies Act and the International

Emergency Economic Powers Act, to curtail trade and transactions between the United States and North Korea. Congress authorizes the President to fine-tune the relationship with North Korea for foreign policy and national security reasons with each waiver authority it incorporates into legislation. At the same time, Congress closely influences the President's choices by enacting issue-driven legislation—addressing human rights matters or proliferation concerns, for example—and by adopting North Korea-specific statutes—most particularly the

- North Korean Human Rights Act of 2004 (P.L. 108-333),
- North Korean Human Rights Reauthorization Act of 2008 (P.L. 110-346),
- North Korea Sanctions and Policy Enhancement Act of 2016 (P.L. 114-122),
- Korean Interdiction and Modernization of Sanctions Act (title III of the Countering America's Adversaries Through Sanctions Act; P.L. 115-44), and
- inclusion of North Korea into the Iran, North Korea, and Syria Nonproliferation Act of 2000 (P.L. 106-178).

As Congress and the President consider proposals to reform foreign aid, streamline export controls, fund defense and international programs, keep proliferation regimes relevant, assess and enter into treaties and international agreements, and participate in multilateral fora, the effectiveness of economic sanctions as a foreign policy and national security tool is likely to be considered. U.S. policy toward North Korea, expressed both unilaterally and in the United States' position in multilateral fora, is further complicated by other considerations—not the least of which include relations with other states in the region, security responsibilities with South Korea and Japan, trade with China, a determination to keep key stakeholders engaged in nonproliferation efforts in both North Korea and elsewhere, and finding the means to balance all U.S. foreign policy and national security interests in a meaningful way.

Appendix A. North Korea—Economic Sanctions Currently Imposed in Furtherance of U.S. Foreign Policy or National Security Objectives

Rationale	Restriction	Statutory basis [regulation]	Authority to impose	Authority to lift or waive
Comprehensive foreign policy reasons, including weapons proliferation, sanctions evasion, violation of UNSC resolutions, censorship, undermining cybersecurity	Block assets, prohibit transactions	Sec. 104, NKSPEA (22 U.S.C. 9214)	President	Mandatory designations in Sec. 104(a); discretionary designations in Sec. 104(b) President may waive for 30 days to 1 year for humanitarian purposes. President may waive for 30 days to 1 year if it is important to national security interests, furthers the enforcement of the act, or "is for an important law enforcement purpose." Sec. 401, NKSPEA (22 U.S.C. 9251) authorizes President to suspend sanctions for up to one year, renewable in 180-day increments. Sec. 402, NKSPEA (22 U.S.C. 9252) authorizes President to terminate sanctions if DPRK dismantles its nuclear and missile programs and resolves human rights issues.
Comprehensive foreign policy reasons	Prohibit indirect correspondent accounts with U.S. financial institutions	Sec. 201A, NKSPEA (22 U.S.C. 9221a)	Related to Sec. 104 designations	Related to Sec. 104 designations President may waive for 30 days to 1 year for humanitarian purposes. President may waive for 30 days to 1 year if it is important to national security interests, furthers the enforcement of the act, or "is for an important law enforcement purpose."

Rationale	Restriction	Statutory basis [regulation]	Authority to impose	Authority to lift or waive
				Sec. 401, NKSPEA (22 U.S.C. 9251) authorizes President to suspend sanctions for up to one year, renewable in 180-day increments. Sec. 402, NKSPEA (22 U.S.C. 9252) authorizes President to terminate sanctions if DPRK dismantles its nuclear and missile programs and resolves human rights issues.
Comprehensive foreign policy reasons	Prohibits U.S. executive agency from entering into procurement contracts with Sec. 104 designees	Sec. 204, NKSPEA (22 U.S.C. 9224)	Related to Sec. 104 designations; Administrator of General Services	Related to Sec. 104 designations President may waive for 30 days to 1 year if it is important to national security interests, furthers the enforcement of the act, or "is for an important law enforcement purpose." Sec. 401, NKSPEA (22 U.S.C. 9251) authorizes President to suspend sanctions for up to one year, renewable in 180-day increments. Sec. 402, NKSPEA (22 U.S.C. 9252) authorizes President to terminate sanctions if DPRK dismantles its nuclear and missile programs and resolves human rights issues.
Comprehensive foreign policy reasons	Vessel, aircraft, or conveyance may be seized and forfeited	Sec. 206, NKSPEA (22 U.S.C. 9226) (pursuant to chapter 46 of 18 U.S.C. or part V, Title IV, Tariff Act of 1930)	Related to Sec. 104 designations	Related to Sec. 104 designations President may waive for 30 days to 1 year if it is important to national security interests, furthers the enforcement of the act, or "is for an important law enforcement purpose." Sec. 401, NKSPEA (22 U.S.C. 9251) authorizes President to suspend sanctions for up to one year, renewable in 180-day increments.

Appendix A. (Continued)

Rationale	Restriction	Statutory basis [regulation]	Authority to impose	Authority to lift or waive
				Sec. 402, NKSPEA (22 U.S.C. 9252) authorizes President to terminate sanctions if DPRK dismantles its nuclear and missile programs and resolves human rights issues.
Comprehensive foreign policy reasons	Deny entry into the U.S. to designated persons, corporate officers of designees, or principal shareholders of designees	Sec. 206, NKSPEA (22 U.S.C. 9226)	Secretary of State, Secretary of Homeland Security	Related to Sec. 104 designations President may waive for 30 days to 1 year for humanitarian purposes. President may waive for 30 days to 1 year if it is important to national security interests, furthers the enforcement of the act, or "is for an important law enforcement purpose." Sec. 401, NKSPEA (22 U.S.C. 9251) authorizes President to suspend sanctions for up to one year, renewable in 180-day increments. Sec. 402, NKSPEA (22 U.S.C. 9252) authorizes President to terminate sanctions if DPRK dismantles its nuclear and missile programs and resolves human rights issues.
General foreign policy reasons	Limits the export of goods or services	Export Controls Act of 2018 (Title XVII, Subtitle B, Part I, P.L. 115-232; 50 U.S.C. 4811 et seq.)	Secretary of Commerce, in coordination with the Secretaries of State and Defense, generally	President, Secretary of Commerce, generally

Rationale	Restriction	Statutory basis [regulation]	Authority to impose	Authority to lift or waive
		[15 C.F.R. Part 730-774]		
General foreign policy reasons	Limits proportionate share to international organizations which, in turn, expend funds in North Korea	Sec. 307, Foreign Assistance Act of 1961 (P.L. 87-195; 22 U.S.C. 2227)	Statutory requirement	No waiver; exemption for certain IAEA programs
General foreign policy reasons	Prohibits bilateral assistance	Sec. 7007, Department of State, Foreign Operations, and Related Programs Appropriations Act, 2019 (Division F, P.L. 116-6; 133 Stat. 13)	Statutory requirement	No waiver
General foreign policy reasons	Prohibits Economic Support Funds	Sec. 7043(c), Department of State, Foreign Operations, and Related Programs Appropriations Act, 2019 (Division F, P.L. 116-6; 133 Stat. 13)	Statutory requirement	No waiver
General foreign policy reasons	Prohibits DOD funds	Sec. 8043, Department of Defense Appropriations, 2019 (Division A, P.L. 115-245; 132 Stat. 3011).	Statutory requirement	No waiver
Diplomatic relations severed	Prohibits most foreign aid and agricultural sales under P.L. 480	Sec. 620(t), Foreign Assistance Act of 1961 (P.L. 87-195; 22 U.S.C. 2370(t))	Statutory requirement	No waiver
National security controls, Communism	Limits the export of goods or services	Export Controls Act of 2018 (Title XVII, Subtitle B, Part I, P.L. 115-232; 50 U.S.C. 4811 et seq.) [15 C.F.R. Part 730-774]	President	President
Communism	Prohibits foreign aid	Sec. 620(f), Foreign Assistance Act of 1961 (P.L. 87-195; 22 U.S.C. 2370(f))	Statutory requirement	President
Communism	Limits proportionate share to international organizations which, in turn, expend funds in North Korea	Sec. 307, Foreign Assistance Act of 1961 (P.L. 87-195; 22 U.S.C. 2227)	Statutory requirement	No waiver; exemption for certain IAEA programs

Appendix A. (Continued)

Rationale	Restriction	Statutory basis [regulation]	Authority to impose	Authority to lift or waive
Communism	Prohibits Export-Import Bank funding to Marxist-Leninist states	Sec. 2(b)(2), Export-Import Bank Act of 1945 (P.L. 79-173; 12 U.S.C. 635(b)(2))	Statutory requirement	President
Communism	Prohibits support in the IFIs	Sec. 43, Bretton Woods Agreements Act (P.L. 79-171; 22 U.S.C. 286aa)	Statutory requirement	Secretary of the Treasury
Communism	Denies favorable trade terms	Sec. 401, Trade Act of 1974 (19 U.S.C. 2431)	Statutory requirement	President
Nonmarket economy and emigration	Denies favorable trade terms	Sec. 402, Trade Act of 1974 (19 U.S.C. 2432)	Statutory requirement	President
Nonmarket economy and emigration	Denies favorable trade terms	Sec. 409, Trade Act of 1974 (19 U.S.C. 2439)	President	President
Communism and market disruption	Denies favorable trade terms	Sec. 406, Trade Act of 1974 (19 U.S.C. 2436)	President	President
Terrorism. Communism	Prohibits the acquisition of property in U.S. for diplomatic mission	Sec. 205, State Department Basic Authorities Act (P.L. 84- 885; 22 U.S.C. 4305)	Secretary of State	Secretary of State
Cybersecurity	Blocks assets, prohibits transactions	Sec. 209, NKSPEA (22 U.S.C. 9229)	President	President may waive for 30 days to 1 year if it is important to national security interests, furthers the enforcement of the act, or "is for an important law enforcement purpose."

Rationale	Restriction	Statutory basis [regulation]	Authority to impose	Authority to lift or waive
				Sec. 401, NKSPEA (22 U.S.C. 9251) authorizes President to suspend sanctions for up to one year, renewable in 180-day increments. Sec. 402, NKSPEA (22 U.S.C. 9252) authorizes President to terminate sanctions if DPRK dismantles its nuclear and missile programs and resolves human rights issues
Cybersecurity	Blocks assets, prohibits transactions	Sec. 210, NKSPEA (22 U.S.C. 9230) [related to Executive Order 13687 and EO 13694, as amended]	President	President Sec. 401, NKSPEA (22 U.S.C. 9251) authorizes President to suspend sanctions for up to one year, renewable in 180-day increments. Sec. 402, NKSPEA (22 U.S.C. 9252) authorizes President to terminate sanctions if DPRK dismantles its nuclear and missile programs and resolves human rights issues.
Cybersecurity	Blocks assets, prohibits transactions	Executive Order 13694 of April 1, 2015, as amended by EO 13757 of December 29, 2016	President	President
Human rights	Prohibit imports of goods made by DPRK prison labor	Sec. 302A, NKSPEA (22 U.S.C. 9241a) [related to Sec. 307, Tariff Act of 1930; 19 U.S.C. 1307]	Commissioner of U.S. Customs and Border Protection	President may waive for 30 days to 1 year if it is important to national security interests, furthers the enforcement of the act, or "is for an important law enforcement purpose." Sec. 401, NKSPEA (22 U.S.C. 9251) authorizes President to suspend sanctions for up to one year, renewable in 180-day increments. Sec. 402, NKSPEA (22 U.S.C. 9252) authorizes President to terminate sanctions if DPRK dismantles its nuclear and missile programs and resolves human rights issues.

Appendix A. (Continued)

Rationale	Restriction	Statutory basis [regulation]	Authority to impose	Authority to lift or waive
Human rights	Blocks assets, prohibitions transactions	Sec. 302B, NKSPEA (22 U.S.C. 9241b) [related to Sec. 302A, NKSPEA]	President	President may make exception for labor meeting international standards. President may waive for 30 days to 1 year if it is important to national security interests, furthers the enforcement of the act, or "is for an important law enforcement purpose." Sec. 401, NKSPEA (22 U.S.C. 9251) authorizes President to suspend sanctions for up to one year, renewable in 180-day increments. Sec. 402, NKSPEA (22 U.S.C. 9252) authorizes President to terminate sanctions if DPRK dismantles its nuclear and missile programs and resolves human rights issues.
Human rights (censorship)	Blocks assets, prohibits transactions	Sec. 304, NKSPEA (22 U.S.C. 9243) [related to Sec. 302A, NKSPEA]	President, related to Sec. 104, NKSPEA	President President may waive for 30 days to 1 year if it is important to national security interests, furthers the enforcement of the act, or "is for an important law enforcement purpose." Sec. 401, NKSPEA (22 U.S.C. 9251) authorizes President to suspend sanctions for up to one year, renewable in 180-day increments. Sec. 402, NKSPEA (22 U.S.C. 9252) authorizes President to terminate sanctions if DPRK dismantles its nuclear and missile programs and resolves human rights issues.
Terrorism	Limits the export of goods and services	Sec. 1754(c), Export Controls Act of 2018 (Title XVII, Subtitle B, Part I, P.L. 115-232; 50 U.S.C. 4811 et seq.)	Secretary of State	Secretary of State, after the President notifies Congress

Rationale	Restriction	Statutory basis [regulation]	Authority to impose	Authority to lift or waive
Terrorism	Prohibits licenses for arms exports/imports	Sec. 38, Arms Export Control Act (P.L. 90-629; 22 U.S.C. 2778) [22 C.F.R. Part 126.1; 27 C.F.R. Part 447.52]	Secretary of State	Secretary of State
Terrorism	Prohibits transactions related to defense articles and defense services	Sec. 40, Arms Export Control Act (P.L. 90-629; 22 U.S.C. 2780)	Secretary of State	Secretary of State, after the President notifies Congress. President may waive on a case- by-case basis. Congress may block a rescission by joint resolution.
Terrorism, failure to cooperate with U.S. efforts	Prohibits transactions related to defense articles and defense services	Sec. 40A, Arms Export Control Act (P.L. 90-629; 22 U.S.C. 2781)	President	President, at annual review, or waived by the President if he finds it "important to the national interests of the United States."
Terrorism	Prohibits most aid under the Foreign Assistance Act of 1961, Agricultural Trade Development and Assistance Act of 1954, Peace Corps Act, and Export-Import Bank Act of 1945	Sec. 620A, Foreign Assistance Act of 1961 (P.L. 87-195; 22 U.S.C. 2371)	Secretary of State	Secretary of State, after the President notifies Congress. President may waive if he finds "that national security interests of humanitarian reasons justify a waiver."
Terrorism	Prohibits imports	Sec. 505, International Security and Development Cooperation Act of 1985 (P.L. 99-83; 22 U.S.C. 2349aa-9)	President	President
Terrorism	Denies Export-Import Bank financing	Sec. 2(b)(1)(B), Export-Import Bank Act of 1945 (P.L. 79-173; 12 U.S.C. 635(b)(1)(B))	President	President

Appendix A. (Continued)

Rationale	Restriction	Statutory basis [regulation]	Authority to impose	Authority to lift or waive
Terrorism	Opposes loans or funding through international financial institutions	Sec. 1621, International Financial Institutions Act (P.L. 95-118; 22 U.S.C. 262p-4q)	Secretary of the Treasury, if a country is listed under §6(j), EAA, or §620A, FAA'61.	Secretary of the Treasury (no waiver authority)
Terrorism	Opposes loans or funding through the International Monetary Fund	Sec. 6, Bretton Woods Agreements Act Amendments of 1978 (P.L. 95-435; 22 U.S.C. 286e-11)	Secretary of the Treasury, if a country is listed under §6(j), EAA, or §620A, FAA'61.	Secretary of the Treasury (no waiver authority)
Terrorism	Prohibits bilateral assistance	Sec. 7021, Department of State, Foreign Operations, and Related Programs Appropriations, 2019 (Division F, P.L. 116-6)	Statutory requirement	No waiver
Terrorism	Limits export licensing for food and medicine; prohibits government financing for such exports	Secs. 906, 908 Trade Sanctions Reform Act of 2000 (P.L. 106-387; 22 U.S.C. 7205, 7207)	Statutory requirement	President, based on national security interests or humanitarian reasons
Terrorism	Limits provision of services to security forces, law enforcement, military, intelligence community	Sec. 40, State Department Basic Authorities Act (P.L. 84- 885; 22 U.S.C. 2712)	Secretary of State	Secretary of State
Terrorism, excessive military expenditure, human rights violations	Prohibits the cancellation or reduction of certain debt	Sec. 501, Miscellaneous Appropriations, 2000 (H.R. 3425, enacted by reference in P.L. 106-113; 22 U.S.C. 2395a note)	Statutory requirement	President

Rationale	Restriction	Statutory basis [regulation]	Authority to impose	Authority to lift or waive
National emergency, proliferation of weapons of mass destruction	Blocks assets of named proliferators of weapons of mass destruction	International Emergency Economic Powers Act (P.L. 95- 223; esp. at 50 U.S.C. 1702); National Emergencies Act (P.L. 94-412; 50 U.S.C. 1601 et seq.)	President [Executive Order 13382, June 28, 2005; 50 U.S.C. 1701 note]	President
National emergency	Prohibits imports, exports, transactions related to transportation	International Emergency Economic Powers Act (P.L. 95- 223; esp. at 50 U.S.C. 1702); National Emergencies Act (P.L. 94-412; 50 U.S.C. 1601 et seq.) [31 C.F.R. Part 510]	President [Executive Order 13466, June 26, 2008; 50 U.S.C. 1701 note]	President
National emergency, proliferation of weapons of mass destruction, attack of the Cheonan, nuclear detonations, missile launches, violation of UNSCR resolutions, counterfeiting of goods and currency, money laundering, smuggling, narcotics trafficking, destabilizing the region	Blocks assets of, and transactions with or on behalf of, named entities	International Emergency Economic Powers Act (P.L. 95- 223; esp. at 50 U.S.C. 1702); National Emergencies Act (P.L. 94-412; 50 U.S.C. 1601 et seq.) Sec. 5, United Nations Participation Act of 1945 (P.L. 79-264; 22 U.S.C. 287c) [31 C.F.R. Part 510]	President [Executive Order 13551, August 30, 2010; 50 U.S.C. 1701 note] [expands on the national emergency declared in E.O. 13466]	President
National emergency, to ensure implementation of import restrictions agreed to in the U.N. Security Council	"Except to the extent provided in statutes or in licenses, regulations, orders, or directives that may be issued pursuant to this order…" prohibits "importation into the United States, directly or indirectly, of	International Emergency Economic Powers Act (P.L. 95- 223; esp. at 50 U.S.C. 1702); National Emergencies Act (P.L. 94-412; 50 U.S.C. 1601 et seq.) Sec. 5, United Nations Participation Act of 1945 (P.L. 79-264; 22	President [Executive Order 13570, April 18, 2011; 50 U.S.C. 1701 note] [expands on the national emergency declared in E.O. 13466]	President

Appendix A. (Continued)

Rationale	Restriction	Statutory basis [regulation]	Authority to impose	Authority to lift or waive
	any goods, services, or technology from North Korea"	U.S.C. 287c)		
National emergency, provocative, destabilizing, and repressive actions of DPRK, including cyber-related actions; violation of UNSC resolutions; human rights abuses	Prohibits transactions with and blocks assets of any entity of the Government of North Korea or the Workers' Party of Korea	International Emergency Economic Powers Act (P.L. 95- 223; esp. at 50 U.S.C. 1702); National Emergencies Act (P.L. 94-412; 50 U.S.C. 1601 et seq.) Immigration and Nationality Act of 1952 (8 U.S.C. 1182(f))	President [Executive Order 13687 of January 2, 2015; 50 U.S.C. 1701 note] [expands on the national emergency declared in E.O. 13466]	President
National emergency, nuclear weapons and missile programs	Prohibits transactions with and blocks assets of any entity of the Government of North Korea or the Workers' Party of Korea. Prohibits transactions with and blocks assets of any person found to engage in or facilitate any of the following activities to the benefit of the Government of North Korea or the Workers' Party of North Korea: (i) operate an entity in North Korea's transportation, mining, energy, or financial services sector; (ii) transactions related to metal, graphite, coal, or software; (iii) human rights	International Emergency Economic Powers Act (P.L. 95- 223; esp. at 50 U.S.C. 1702); National Emergencies Act (P.L. 94-412; 50 U.S.C. 1601 et seq.) Sec. 5, United Nations Participation Act of 1945 (P.L. 79-264; 22 U.S.C. 287c)	President [Executive Order 13722 of March 15, 2016; 50 U.S.C. 1701 note] [expands on the national emergency declared in E.O. 13466]	President

Rationale	Restriction	Statutory basis [regulation]	Authority to impose	Authority to lift or waive
	violations; (iv) exporting of workers from North Korea; (v) significant activities undermining cybersecurity; (vi) censorship; (vii) materially supported a blocked person or entity; (viii) owned by a blocked person or entity; and (ix) attempt to engage in any of the above restricted activities. Prohibits all exportation from the United States to North Korea. Prohibits new investment in North Korea by a U.S. person. Prohibits financing or guarantee by a U.S. person for a foreign person to engage in any of the above-restricted activities.	North Korea Sanctions and Policy Enhancement Act of 2016 (P.L. 114-122; 22 U.S.C. 9201 et seq.) Immigration and Nationality Act of 1952 (8 U.S.C. 1182(f))	["in view of UNSC Resolution 2270 of March 2, 2016"]	
National emergency, nuclear weapons and missile programs	Authorizes the Treasury Secretary to designate for sanctions anyone who operates in "construction, energy, financial services, fishing, information technology, manufacturing, medical, mining, textiles, or transportation industries in North Korea"; owns or controls a DPRK port—sea, air, or land; Engages "in at least one significant importation from or exportation to North Korea"; Is a North Korean person; Is a North Korean person who raises revenues for the DPRK government or the Workers' Party; Evades sanctions. Prohibits any aircraft or vessel (that has any foreign ownership) that has landed or entered a port in the DPRK to enter the U.S. within 180 days. Fully blocks any funds that arrive in the U.S. that	International Emergency Economic Powers Act (P.L. 95-223; esp. at 50 U.S.C. 1702); National Emergencies Act (P.L. 94-412; 50 U.S.C. 1601 et seq.) Sec. 5, United Nations Participation Act of 1945 (P.L. 79-264; 22 U.S.C. 287c) Immigration and Nationality Act of 1952 (8 U.S.C. 1182(f))	President [Executive Order 13810 of September 20, 2017; 50 U.S.C. 1701 note] [expands on the national emergency declared in E.O. 13466] ["in view of UNSC Resolution 2321 of November 30, 2016, UNSCR 2356 of June 2, 2017, UNSCR 2371 of August 5, 2017, and UNSCR 2375 of September 11, 2017"]	President

Appendix A. (Continued)

Rationale	Restriction	Statutory basis [regulation]	Authority to impose	Authority to lift or waive
	originate from, are destined for, or pass through, a foreign account that is owned or controlled by a DPRK person, or any account that has been used to transfer funds in which a DPRK person has an interest. Prohibits a US person from facilitating in any way such a transaction. Targets foreign financial institutions—those that conduct transactions on behalf of anyone designated for sanctions, transactions related to DPRK trade. Such institutions are denied access to the U.S. banking system (no correspondent accounts or payable-through accounts), their U.S.-based assets and property are blocked.			
Proliferation of weapons of mass destruction: missiles	Restricts export licenses, withholds foreign assistance	Sec. 203, NKSPEA (22 U.S.C. 9223)	Department of Commerce, President, Secretary of State	Secretary of State, if in the national interest Sec. 401, NKSPEA (22 U.S.C. 9251) authorizes President to suspend sanctions for up to one year, renewable in 180-day increments. Sec. 402, NKSPEA (22 U.S.C. 9252) authorizes President to terminate sanctions if DPRK dismantles its nuclear and missile programs and resolves human rights

				issues.
Rationale	**Restriction**	**Statutory basis [regulation]**	**Authority to impose**	**Authority to lift or waive**
Proliferation of weapons of mass destruction: missiles	Prohibits entry into U.S. waters or ports of any vessel that has, within 180 days, entered into port of designated territories (including DPRK)	Sec. 16, Ports and Waterways Safety Act (33 U.S.C. 1232c, as added by Sec. 315(a) of the KIMS Act)	Secretary of State	No waiver
Proliferation of weapons of mass destruction: missiles	Prohibits a range of transactions—U.S. Government contracts, export licenses, imports into United States	Sec. 73, Arms Export Control Act (P.L. 90-629; 22 U.S.C. 2797b)	President	President
Proliferation of weapons of mass destruction: nuclear enrichment transfers	Prohibits foreign aid, military aid	Sec. 101, Arms Export Control Act (P.L. 90-629; 22 U.S.C. 2799aa)	President	President
Proliferation of weapons of mass destruction: nuclear reprocessing transfers, nuclear detonations	Prohibits foreign aid (except humanitarian), military aid, USG defense sales and transfers, export licenses for USML goods and services, U.S. Government-backed credits, support in the international banks, agricultural credits or financing, U.S. commercial bank financing, licenses for export of certain goods and services	Sec. 102, Arms Export Control Act (P.L. 90-629; 22 U.S.C. 2799aa-1)	President	President
Proliferation of weapons of mass destruction: nuclear detonations	Prohibits Export-Import Bank financing	Sec. 2(b)(4) of the Export- Import Bank Act of 1945 (P.L. 79-173; 12 U.S.C. 635(b)(4))	Statutory requirement	President

Appendix A. (Continued)

Rationale	Restriction	Statutory basis [regulation]	Authority to impose	Authority to lift or waive
Proliferation of weapons of mass destruction: nuclear detonations	Prohibits Export-Import Bank financing	Title VI of the Department of State, Foreign Operations, and Related Programs Appropriations Act, 2019 (Division F, P.L. 116-6; 133 Stat. 13)	Statutory requirement	No waiver
Proliferation of weapons of mass destruction: missiles	Prohibits a range of transactions—contracts, export licenses, imports into U.S.	Sec. 11B, Export Administration Act (P.L. 96-72; 50 U.S.C. App. 2410b), as continued by Sec. 1766, Export Controls Act of 2018 (Part I, Subtitle B, Title XVII, P.L. 115-232), requiring annual continuation under IEEPA authorities	President	President
Proliferation of weapons of mass destruction	Prohibits a range of transactions—arms sales and exports, dual-use exports, procurement contracts, assistance, imports, support in the international banks, credit, air landing rights	Sec. 3, Iran, North Korea, and Syria Nonproliferation Act of 2000 (P.L. 106-178; 50 U.S.C. 1701 note)	President	President
Use of chemical or biological weapons	In two rounds, prohibits (I) foreign assistance, arms sales, USML/CCL export licenses, foreign military financing, USG credits or financial assistance; and (2) support in the IFIs, US bank loans and credits, wider	Chemical and Biological Weapons Control and Warfare Elimination Act of 1991 (CBW Act; Title III, P.L. 102-1821 22 U.S.C. 5601 et seq.).	President (delegated to Secretary of State)	President

Rationale	Restriction	Statutory basis [regulation]	Authority to impose	Authority to lift or waive
	exports, imports, normal diplomatic relations, air landing rights			
Human rights (trafficking in persons)	Prohibits nonhumanitarian foreign aid, cultural exchanges, support in international financial institutions	Sec. 110, Trafficking Victims Protection Act of 2000 (P.L. 106-386; 22 U.S.C. 7107)	President	President, waiver if in the national interest
Counterfeiting, money- laundering	Prohibits certain commercial bank transactions	31 U.S.C. 5318A (generally referred to by its amendatory vehicle—Sec. 311, USA PATRIOT Act) (See also Sec. 201(c), NKSPEA)	Secretary of the Treasury (delegated to Financial Crimes Enforcement Network (FinCEN), which declared DPRK as a "primary jurisdiction of money laundering concern" effective June 2, 2016 (final rule entered into effect December 9, 2016).	Secretary of the Treasury

Source: Congressional Research Service.

Notes: Most of the Export Administration Act of 1979 (P.L. 96-72) was repealed with the enactment of the Export Controls Act of 2018 (ECA'18; Title XVII, Subtitle B, Part I, P.L. 115-232; 50 U.S.C. 4811 et seq.). This affects several aspects of the sanctions regime imposed on North Korea, including export controls established for foreign policy and national security reasons, and terrorism designations. Section 1768(a) of the ECA'18 (50 U.S.C. 4826) provides that:

All delegations, rules, regulations, orders, determinations, licenses, or other forms of administrative action that have been made, issued, conducted, or allowed to become effective under the Export Administration Act of 1979 (50 U.S.C. 4601 et seq.) (as in effect on the day before the enactment of this Act and as continued in effect pursuant to the International Emergency Economic Powers Act (50 U.S.C. 1701 et seq.)), or the Export Administration Regulations, and are in effect as of the date of the enactment of this Act [August 13, 2018], shall continue in effect according to their terms until modified, superseded, set aside, or revoked under the authority of this part.

NKSPEA = North Korea Sanctions and Policy Enhancement Act of 2016 (P.L. 114-122; 22 U.S.C. 9201 et seq.), as amended; USML = United States Munitions List; CCL = Commerce Control List; IEEPA = International Emergency Economic Powers Act; USG = United States Government; IFI = International Financial Institutions.

Appendix B. DPRK Sanctions: Comparison of U.N. Security Council Requirements and U.S. Sanctions

	U.N. Security Council Resolution Requirements	U.S. Law	Executive Order
Financial Services	UNSCR 1874 (2009) prohibits providing financial services that could contribute to DPRK's nuclear-, missile-, or WMD-related programs.	Sec. 104(b)(1)(F), NKSPEA, as amended, authorizes (but does not require) the President to designate any person who transfers funds in violation of UNSC requirements.	EO 13466 (2008) prohibits any transaction by a U.S. person the purpose of which is to evade sanctions. EO 13551 (2010) blocks the assets of specific DPRK individuals and entities, pursuant to emergency authorities and requirements of UNSCRs 1718 (2006) and 1874 (2009). Also blocks assets of anyone found to: --engage in arms trade with DPRK; --provide training, advice, or other services related to arms manufacture, maintenance, or use; --trade in luxury goods; --engage in money laundering, counterfeiting of goods or currency, trafficked in bulk cash, narcotics or other illicit economic activity; or --engage in sanctions evasion.
	UNSCR 1874 (2009) "calls on" member states not to enter into new financial commitments for aid or loans, including "public financial support for trade" to DPRK.		
	UNSCR 2087 (2013) "calls on" member states to exercise vigilance on those within their jurisdictions working on behalf of DPRK financial institutions and their branches, representatives, and subsidiaries.	Sec. 104(b)(1)(N), NKSPEA, as amended, authorizes (but does not require) the President to designate any person who facilitates operations of any DPRK financial institution.	
	UNSCR 2087 (2013) "calls on" member states to exercise vigilance relating to the use of bulk cash to evade sanctions.	Sec. 201(b), NKSPEA requires the President to determine whether DPRK is a "jurisdiction of primary money laundering concern." [Such a determination was made in June 2016.]	
	UNSCR 2094 (2013) expands the list of prohibited military-use goods, and prohibits "brokering or intermediary services, including...arranging for the provision, maintenance, or use of prohibited items...or the supply, sale, or transfer to or exports from other states."		
	UNSCR 2094 (2013) "calls on" member states to prohibit financial institutions from opening offices or subsidiaries in DPRK if such offices would benefit DPRK's weapons programs.	Sec. 104(a)(6), NKSPEA, requires President to designate any person who engages in bulk money laundering, bulk cash smuggling, or narcotics trafficking that supports the DPRK government.	

	U.N. Security Council Resolution Requirements	U.S. Law	Executive Order
Financial Services	UNSCR 2094 (2013) prohibits providing "public financial support for trade" to DPRK if such aid would benefit DPRK's weapons programs. UNSCR 2270 (2016) prohibits the opening and operation of new branches, subsidiaries, and representative offices of DPRK banks. UNSCR 2270 (2016) prohibits new joint ventures, ownership, or correspondent relationships with DPRK banks. UNSCR 2270 (2016) requires member states to terminate joint ventures, ownership, or correspondent relationships with, and close existing branches, subsidiaries, and representative offices of, DPRK banks. UNSCR 2270 (2016) prohibits the opening of new representative offices, branches, or banking accounts in DPRK. UNSCR 2270 (2016) requires the closing of existing representative offices, branches, or banking accounts in DPRK, if there is credible information that the accounts and services contribute to DPRK's weapons programs. UNSCR 2270 (2016) prohibits public and private financial support for trade with DPRK (e.g., export credits, guarantees, insurance) where such support could contribute to DPRK's weapons programs. UNSCR 2270 (2016) prohibits transfers of gold to or from DPRK.	Sec. 104(b)(1)(G), NKSPEA, as amended, authorizes (but does not require) the President to designate any person who facilitates transfers bulk cash, precious metals, gems, other valuables not already covered. Sec. 104(a)(14), NKSPEA, as amended, requires President to designate any person who maintains a correspondent account for a DPRK financial institution. Sec. 201A, NKSPEA, as amended, prohibits indirect correspondent accounts.	EO 13722 (2016) blocks property in the U.S. of and person or entity that operates in any industry in the DPRK economy, "such as transportation, mining, energy, or financial services." In addition, EO 13722 blocks assets of anyone who: --trades in metal, graphite, coal, or software, with the government of DPRK or the Workers' Party; --engages in or facilitates human rights abuses in DPRK; --exports labor from DPRK; --engages in or facilitates undermining cybersecurity on behalf of DPRK; --engages in or facilitates censorship; --engages in sanctions evasion; or --attempts to engage in any of these behaviors.

Appendix B. (Continued)

	U.N. Security Council Resolution Requirements	U.S. Law	Executive Order
Financial Services	UNSCR 2270 (2016) "calls on" member states to use targeted sanctions to curtail the financing of DPRK's weapons programs, following guidelines of the Financial Action Task Force (FATF).a UNSCR 2321 (2016) requires member states to limit bank accounts of DPRK diplomatic missions to one, and one per accredited diplomat and consular officer. UNSCR 2321 (2016) prohibits DPRK from using consular property in member states for other than diplomatic or consular activities (no commerce permitted). UNSCR 2321 (2016) prohibits the providing of insurance to DPRK vessels.		In addition, the Order prohibits all export from the U.S. to DPRK; new investment in DPRK; and financing, facilitation, or guarantee of a transaction by a US person for a prohibited transaction or designee. EO 13810 of September 20,
	UNSCR 2397 (2017) prohibits the providing of insurance to any DPRK vessel for which there are reasonable grounds that it is violating UNSC requirements. UNSCR 2321 (2016) requires member states to close representative offices, subsidiaries, and banking accounts in DPRK..		2017, blocks any funds that arrive in the US that originate from, are destined for, or pass through, a foreign account that is owned or controlled by a DPRK person, or any account that has been used to transfer funds in which a DPRK person has an interest. Prohibits a US person from facilitating in any way such a transaction.

	U.N. Security Council Resolution Requirements	U.S. Law	Executive Order
Financial Services	UNSCR 2321 (2016) prohibits private and public financing for trade within DPRK. UNSCR 2321 (2016) "calls on" member states to be alert to bulk cash used to evade sanctions. UNSCR 2371 (2017) prohibits the opening of new, or expanding existing, joint ventures or cooperative entities with DPRK entities or individuals. UNSCR 2371 (2017) prohibits the use of financial services to clear funds for DPRK entities. Broadens the application of the term "banks" to include any entity that provides a financial service		EO 13810 of September 20, 2017, targets foreign financial institutions— those that conduct transactions on behalf of anyone designated for sanctions, transactions related to DPRK trade. Such institutions are denied access to the U.S. banking system (no correspondent accounts or payable-through accounts), their U.S.-based assets and property are blocked
Arms and Related Materiel Embargo	UNSCR 1718 (2006) prohibits supply, sale, or transfer to DPRK of battle tanks, armored combat vehicles, artillery systems, combat aircraft, attack helicopters, warships, missiles, or missile systems, or related spare parts. The measure also prohibits DPRK from exporting such goods. UNSCR 1874 (2009) expands the above list of prohibited goods, and to prohibit financial transactions related to such trade. UNSCR 2087 (2013) expands the above list of prohibited goods.	Secs. 2, 38, 40, 42, 71, AECA (general authorities governing arms trade) Sec. 104(a)(1), NKSPEA, requires President to designate any person who trades in goods or services that	The U.S. maintains a prohibition on trading in arms and related material with DPRK in regulations— 22 C.F.R. Part 120-130 (See, especially, 22 C.F.R. Part 126; State Department International Traffic in Arms

Appendix B. (Continued)

	U.N. Security Council Resolution Requirements	U.S. Law	Executive Order
Arms and Related Materiel Embargo	UNSCR 2094 (2013) expands the above list of prohibited goods. UNSCR 2270 (2016) expands the above list of prohibited goods. UNSCR 2321 (2016) expands the above list of prohibited goods. UNSCR 2321 (2016) requires the Sanctions Committee to adopt a "conventional arms dual-use list" for prohibited goods. UNSCR 2371 (2017) requires the Sanctions Committee to expand the list of prohibited goods, including additional conventional arms-related items. UNSCR 2375 (2017) expands the list of prohibited goods to include additional dual- use goods, conventional arms-related items and other materials.	benefit DPRK's weapons programs. Sec. 104(a)(9), NKSPEA, requires President to designate any person who trades in arms or related materiel to or from DPRK. Sec. 203(a), NKSPEA, requires DPRK to be treated as a state sponsor of international terrorism to the extent it impedes the export of U.S. dual-use goods and services, and munitions. Sec. 203(b), NKSPEA, as amended, requires the President to withhold foreign aid to third countries that provide DPRK defense articles and defense services.	Regulations), based primarily on AECA authorities; and 15 C.F.R. Part 700-799 (see, especially, 15 C.F.R. Part 746.4; Department of Commerce Export Administration Regulations), based primarily on EAA. EO 13551 (2010) blocks the assets of specific DPRK individuals and entities, pursuant to emergency authorities and requirements of UNSCRs 1718 (2006) and 1874 (2009). Also blocks assets of anyone found to: --engage in arms trade with DPRK; --provide training, advice, or other services related to arms manufacture, maintenance, or use; --trade in luxury goods; --engage in money laundering, counterfeiting of goods or currency, trafficked in bulk cash, narcotics or other illicit economic activity; or --engage in sanctions evasion.

	U.N. Security Council Resolution Requirements	U.S. Law	Executive Order
Non- Proliferation and Proliferation Networks	UNSCR 1718 (2006) prohibits supply, sale, or transfer to DPRK of specific items, materials, equipment, goods, and technology related to nuclear, ballistic missile, or other WMD-related program. The measure also prohibits DPRK from exporting such goods. UNSCR 1874 (2009) expands the above list of prohibited goods and prohibits financial transactions related to such trade. UNSCR 2087 (2013) expands the above list of prohibited goods. UNSCR 2094 (2013) expands the above list of prohibited goods. UNSCR 2270 (2016) expands the above list of prohibited goods. UNSCR 2321 (2016) expands the above list of prohibited goods.	Sec. 2, 3, Iran, North Korea, Syria Non-proliferation Act, authorizes the President to block assets of and transactions with any foreign person designated as having Transferred goods, services, or technology related to certain weapons programs of Iran, Syria, or North Korea.	See arms embargo regulations, above.
Interdiction, Inspection, and Transportation (See also "Helicopters and Vessels (including aircraft, landing and flyover rights, port use)," below)	UNSCR 1718 (2006) "calls on" member states to inspect cargo to/from DPRK, "as necessary." UNSCR 1874 (2009) expands this to "call on" states to inspect all cargo to/from DPRK, including seaports and airports. UNSCR 2094 (2013) requires member states to inspect all cargo transiting their territory that originates in or is destined for DPRK "if the State concerned has credible information" that the cargo is prohibited under earlier resolutions. Any vessel that resists inspection shall be denied entry.	Sec. 104(b)(1)(M), NKSPEA, as amended, authorizes (but does not require) the President to designate any person who engages in transactions in DPRK's transportation, mining, energy, or financial services sectors.	EO 13722 (2016) blocks property in the U.S. of and person or entity that operates in any industry in the DPRK economy, "such as transportation, mining, energy, or financial services." EO 13810 of September 20, 2017, authorizes sanctions on anyone who operates in construction, energy, financial services, fishing, information technology, manufacturing, medical, mining, textiles, or transportation industries in the DPRK.

Appendix B. (Continued)

	U.N. Security Council Resolution Requirements	U.S. Law	Executive Order
Interdiction, Inspection, and Transportation (See also "Helicopters and Vessels (including aircraft, landing and flyover rights, port use)," below)	UNSCR 2094 (2013) "calls on" member states to deny permission to aircraft to take off, land, or fly over if the aircraft is suspected of carrying prohibited cargo. UNSCR 2270 (2016) requires member states to inspect all cargo transiting airports, seaports, and free trade zones that is headed to/from DPRK, or has been brokered by a DPRK national, or is carried on a DPRK-flagged vessel. UNSCR 2321 (2016) expands the definition of "cargo," and thus subject to inspection, to include personal luggage and checked bags. UNSCR 2321 (2016) adds "rail" and "road" to the list of transit means to be inspected. UNSCR 2371 (2017) authorizes the Sanctions Committee to designate additional vessels to be subject to sanctions—assets freeze, seizure, denial of entry into port. UNSCR 2375 (2017) directs the 1718 Committee to designate vessels transporting prohibited items from DPRK.	Sec. 16, Ports and Waterways Safety Act, as amended by KIMS Act, denies entry into U.S. ports to any foreign vessel designated as noncompliant with UNSC resolutions.	EO 13810 of September 20, 2017 authorizes sanctions on anyone who owns or controls a DPRK port—sea, air, or land.

	U.N. Security Council Resolution Requirements	U.S. Law	Executive Order
Interdiction, Inspection, and Transportation (See also "Helicopters and Vessels (including aircraft, landing and flyover rights, port use)," below)	UNSCR 2397 (2017) requires member states to de-register any vessel for which there are reasonable grounds that it violates UNSC requirements. Prohibits a state from registering a vessel that has been de-registered by another. UNSCR 2397 (2017) requires member states to seize, inspect, and impound any vessel in their jurisdiction for which there are reasonable grounds that it violates UNSC requirements	Sec. 104(a)(12), NKSPEA, as amended, requires President to designate any person who provides	
Bunkering Services (Providing Fuel, Supplies, or Servicing)	UNSCR 1874 (2009) prohibits the providing of bunkering services to vessels suspected of carrying prohibited cargo. UNSCR 2270 (2016) prohibits selling or supplying aviation fuel, gasoline, jet fuels, rocket fuels, to DPRK. UNSCR 2321 (2016) "calls on" member states to provide no more aviation fuel than is required for a "margin for safety of flight." UNSCR 2375 (2017) requires a full ban on the supply, sale, or transfer of all condensates and natural gas liquids to DPRK.	Bunkering services to vessels from DPRK.\ Sec. 104(a)(11), NKSPEA, as amended, requires President to designate any person who trades in various aviation, jet, and rocket fuels to DPRK.	
Assets Freeze	UNSCR 1718 (2006) freezes funds, financial assets, and economic resources of those designated by the Sanctions Committee or Security Council for supporting DPRK's nuclear-, missile-, or WMD-related programs. UNSCR 1874 (2009) authorizes each member state to freeze funds, financial assets, and economic resources if the member state finds them to be supporting DPRK's nuclear-, missile-, or WMD-related programs.	National Emergencies Act; International Emergency Economic Powers Act Sec. 104(a)(4), NKSPEA, requires President to designate any person who engages in censorship by the DPRK government.	EO 13466 (2008) blocks all property and interests in property of DPRK that was blocked as of June 16, 2000, and remained blocked as of June 26, 2008 (in effect, grandfathering in sanctions in place under TWEA authorities).

Appendix B. (Continued)

	U.N. Security Council Resolution Requirements	U.S. Law	Executive Order
Assets Freeze	UNSCR 2094 (2013) requires each member state to freeze funds, financial assets, and economic resources if the member state finds them to be trading in bulk cash, evading sanctions, or supporting DPRK's nuclear-, missile-, or WMD-related programs. UNSCR 2270 (2016) defines assets to include economic resources "of every kind, tangible or intangible, movable or immovable, actual or potential...including vessels...." UNSCR 2270 (2016) requires the closing of representative offices of designated entities. UNSCR 2270 (2016) requires the freezing of assets of the DPRK government or Workers' Party of Korea. UNSCR 2321 (2016) freezes any vessel (as an asset) found by the Sanctions Committee to be engaged in DPRK's weapons programs.	Sec. 104(a)(5), NKSPEA, requires President to designate any person who engages in human rights abuses by the DPRK government. Sec. 104(a)(7), NKSPEA, requires President to designate any person who engages in undermining cybersecurity on behalf of the DPRK government. Sec 209(b) of that Act also requires the President to designate any person who undermines cybersecurity through the use of computer networks or systems against foreign persons, governments, or other entities on behalf of the DPRK government.	EO 13551 (2010) blocks the assets of specific DPRK individuals and entities, pursuant to emergency authorities and requirements of UNSCRs 1718 (2006) and 1874 (2009). Also blocks assets of anyone found to: --engage in arms trade with DPRK; --provide training, advice, or other services related to arms manufacture, maintenance, or use; --trade in luxury goods; --engage in money laundering, counterfeiting of goods or currency, trafficked in bulk cash, narcotics or other illicit economic activity; or --engage in sanctions evasion.

	U.N. Security Council Resolution Requirements	U.S. Law	Executive Order
Assets Freeze		Sec. 104(c), NKSPEA requires the President to block the assets and prohibit transactions with the DPRK government or Workers' Party of Korea. Sec. 304(b), NKSPEA, requires the President to impose sanctions on anyone identified in a report required by the section who engages in human rights abuses or censorship.	EO 13722 (2016) blocks property in the U.S. of and person or entity that operates in any industry in the DPRK economy, "such as transportation, mining, energy, or financial services." In addition, EO 13722 blocks assets of anyone who: --trades in metal, graphite, coal, or software, with the government of DPRK or the Workers' Party; --engages in or facilitates human rights abuses in DPRK; --exports labor from DPRK; --engages in or facilitates undermining cybersecurity on behalf of DPRK; --engages in or facilitates censorship; --engages in sanctions evasion; or --attempts to engage in any of these behaviors.

Appendix B. (Continued)

	U.N. Security Council Resolution Requirements	U.S. Law	Executive Order
Assets Freeze			In addition, the Order prohibits all export from the U.S. to DPRK; new investment in DPRK; and financing, facilitation, or guarantee of a transaction by a US person for a prohibited transaction or designee. EO 13687 (2015) freezes assets in the U.S. of instrumentalities, agencies, controlled entities, or officials of the DPRK government or Workers' Party of Korea. EO 13722 (2016) freezes assets in the U.S. of the DPRK government or Workers' Party of Korea.
Travel Ban (includes diplomatic entry)	UNSCR 1718 (2006) prohibits travel of those designated (and their family members) by the Sanctions Committee or Security Council for supporting DPRK's nuclear-, missile-, or WMD-related programs.	Immigration and Nationality Act (8 U.S.C. 1102). Sec. 206, NKSPEA, authorizes the Secretary of State to deny entry into the U.S. of any designated person, corporate official of a designee, or shareholder of a designee.	

	U.N. Security Council Resolution Requirements	U.S. Law	Executive Order
Travel Ban (includes diplomatic entry)	UNSCR 2094 (2013) applies the visa denial to those evading sanctions, and requires the expulsion of DPRK nationals who are found to be in violation of relevant UNSCRs. UNSCR 2094 (2013) "calls on" member states to prohibit DPRK's diplomats from benefiting from sanctions evasion. UNSCR 2270 (2016) approves of member states expelling DPRK diplomats if they are found to be working on behalf of a designated entity or person, or assisting in sanctions evasion. UNSCR 2270 (2016) approves of member states expelling foreign nationals if they are found to working on behalf of a designated entity or person, or assisting in sanctions evasion. UNSCR 2270 (2016) requires member states to expel DPRK nationals who are representatives of designated entities. UNSCR 2321 (2016) requires member states to reduce DPRK diplomatic staff numbers in their states. UNSCR 2321 (2016) requires member states to restrict travel to or through their states of DPRK government officials, and DPRK military, if such individuals are associated with DPRK's weapons programs. UNSCR 2321 (2016) requires member states to expel any foreign national working for a DPRK bank or financial institution.		[Article 9, Section 1 of the Vienna Convention on Diplomatic Relations authorizes member states to expel foreign diplomats.]b Vienna Convention. Vienna Convention.

Appendix B. (Continued)

	U.N. Security Council Resolution Requirements	U.S. Law	Executive Order
Specialized Training and Teaching	UNSCR 1718 (2006) prohibits the providing of training or assistance to DPRK related to a wide range of arms and material with military applications, specific items, materials, equipment, goods, and technology related to nuclear, ballistic missile, or other WMD-related program. UNSCR 1874 (2009) expands the above list of prohibited goods, and to prohibit financial transactions related to such training or assistance. UNSCR 1874 (2009) "calls on" member states to be vigilant about teaching DPRK nationals in disciplines related to nuclear and missile proliferation. UNSCR 2087 (2013) expands the above list of prohibited goods and prohibits financial transactions related to such training or assistance. UNSCR 2270 (2016) clarifies that the training ban adopted in 2009 includes a prohibition on hosting trainers, advisors, or other officials related to military training. UNSCR 2270 (2016) prohibits specialized teaching or training of DPRK nationals that could contribute to DPRK's WMD programs. Names specific fields of study. UNSCR 2321 (2016) expands the ban on specialized teaching or training.	Sec. 104(a)(2), NKSPEA, requires President to designate any person who provides training, advice, or other services or assistance related to DPRK's weapons programs.	EO 13551 (2010) blocks the assets of specific DPRK individuals and entities, pursuant to emergency authorities and requirements of UNSCRs 1718 (2006) and 1874 (2009). Also blocks assets of anyone found to: --engage in arms trade with DPRK; --provide training, advice, or other services related to arms manufacture, maintenance, or use; --trade in luxury goods; --engage in money laundering, counterfeiting of goods or currency, trafficked in bulk cash, narcotics or other illicit economic activity; or --engage in sanctions evasion.

	U.N. Security Council Resolution Requirements	U.S. Law	Executive Order
Scientific and Technical Cooperation	UNSCR 2321 (2016) suspends scientific and technical cooperation with DPRK except for medical exchanges, unless the Sanctions Committee determines such cooperation does not contribute to DPRK's weapons programs.		
Trade Ban On Natural Resources	UNSCR 2270 (2016), as amended by UNSCR 2321 (2016), and further amended by UNSCR 2371 (2017), prohibits trade in DPRK coal, iron, or iron ore; ends contracts extending beyond 30 days from the adoption of the 2017 resolution; and eliminates the caps and "livelihood purposes" exceptions relating to coal stated in previous resolutions. UNSCR 2270 (2016) prohibits trade in DPRK gold, titanium ore, vanadium ore, and rare earth elements.	Sec. 104(a)(8), NKSPEA, requires President to designated any person who trades in precious metals, graphite, raw metals, aluminum, steel, coal, software, related to DPRK's weapons programs, the Workers' Party, armed forces, internal security or intelligence activities, prison or forced labor. Sec. 104(b)(1)(D), NKSPEA, as amended, authorizes (but does not require) the President to designate any person who trades in coal, iron, or iron ore in excess of UNSC limits. Sec. 104(a)(10), NKSPEA, as amended, requires President to designate any person who trades in copper, vanadium, gold, titanium, silver, nickel, zinc, or rare earth elements with DPRK.	EO 13570 (2011) prohibits importation into the United States from DPRK of any goods, services, or technology. See EO 13722 (2016), above. EO 13810 of September 20, 2017, authorizes sanctions on anyone who engages in "at least one significant importation from or exportation to North Korea...." See EO 13722 (2016), above.

Appendix B. (Continued)

	U.N. Security Council Resolution Requirements	U.S. Law	Executive Order
Trade Ban On Natural Resources	UNSCR 2321 (2016) prohibits trade in DPRK copper, nickel, silver, and zinc. UNSCR 2371 (2017) prohibits trade in DPRK lead and lead ore. UNSCR 2397 (2017) prohibits the purchase of food, agricultural products, machinery, electrical equipment, earth and stone, magnesite and magnesia, wood, and vessels from DPRK. UNSCR 2397 (2017) prohibits the providing to DPRK of industrial machinery, transportation vehicles, iron, steel, other metals (not including spare parts of civil aircraft safety).	Sec. 104(b)(1)(H), NKSPEA, as amended, authorizes (but does not require) the President to designate any person who facilitates transfers of crude oil, petroleum, liquefied natural gas, other natural gas to the DPRK government.	
Trade Ban On Food Products	UNSCR 2371 (2017) prohibits trade in DPRK seafood.	Sec. 104(b)(1)(K), NKSPEA, as amended, authorizes (but does not require) the President to designate any person who purchases food or agricultural products from the DPRK government.	See EO 13570 (2011), above. See EO 13722 (2016), above. EO 13810 of September 20, 2017, authorizes sanctions on anyone who engages in "at least one significant importation from or exportation to North Korea...."
Trade Ban On Statues	UNSCR 2321 (2016) prohibits trade in DPRK-origin statuary.		See EO 13570 (2011), above.

	U.N. Security Council Resolution Requirements	U.S. Law	Executive Order
Trade Ban On Statues			See EO 13722 (2016), above. EO 13810 of September 20, 2017, authorizes sanctions on anyone who engages in "at least one significant importation from or exportation to North Korea...."
Ban On Trade And Use Of Helicopters And Vessels (Including Aircraft, Landing And Flyover Rights, Port Use)	UNSCR 2094 (2013) "calls on" member states to deny permission to aircraft to take off, land, or fly over if the aircraft is suspected of carrying prohibited cargo. UNSCR 2270 (2016) prohibits the leasing or chartering of vessels or aircraft, or providing crew services, to DPRK, designated entities and individuals, and anyone found to be engaged in sanctions evasion. UNSCR 2270 (2016) "calls on" member states to revoke vessel registrations of those granted designated entities and individuals or those engaged in sanctions evasion; and "calls on" member states not to register a vessel that has been deregistered by another state. UNSCR 2270 (2016) prohibits registering a vessel in DPRK or using a DPRK flag for a foreign vessel. UNSCR 2270 (2016) prohibits aircraft landing and flyover rights to DPRK aircraft.	Sec. 104(a)(13), NKSPEA, as amended, requires President to designate any person who registers a vessel owned by the government of DPRK.	See EO 13570 (2011), above. See EO 13722 (2016), above. EO 13810 of September 20, 2017, authorizes sanctions on anyone who engages in "at least one significant importation from or exportation to North Korea...." EO 13466 (2008) prohibits U.S. persons from registering a vessel in DPRK, using the DPRK flag for a vessel, or owning, leasing, operating, or insuring a DPRK-flagged vessel.

Appendix B. (Continued)

	U.N. Security Council Resolution Requirements	U.S. Law	Executive Order
Ban On Trade And Use Of Helicopters And Vessels (Including Aircraft, Landing And Flyover Rights, Port Use	UNSCR 2270 (2016) prohibits entry into port of DPRK vessels owned or controlled by a designated entity or individual, or contains prohibited cargo. UNSCR 2321 (2016) expands the ban on providing crew services to DPRK to be prohibited without exception. UNSCR 2321 (2016) expands the ban on registering vessels in DPRK to be prohibited without exception. UNSCR 2321 (2016) requires member states to deflag, and deny entry to port, vessels found by the Sanctions Committee to be engaged in DPRK's weapons programs. UNSCR 2321 (2016) requires member states to deregister vessels owned by DPRK, and prohibits the registration of vessels deregistered by a third state. UNSCR 2321 (2016) prohibits trade in DPRK new helicopters and vessels. UNSCR 2375 (2017) authorizes the Sanctions Committee to designate for sanctions vessels transporting prohibited items from DPRK. UNSCR 2397 (2017) prohibits providing new or used vessels to DPRK.		EO 13810 of September 20, 2017, prohibits any aircraft or vessel that is foreign- owned to enter the United States if it has landed or entered a port of North Korea within the last 180 days.

	U.N. Security Council Resolution Requirements	U.S. Law	Executive Order
Ban On Exported Labor	UNSCR 2321 (2016) prohibits the procuring of crew services from DPRK for vessels or aircraft. UNSCR 2321 (2016) "calls on" member states to exercise vigilance over wages paid to DPRK exported labor that may finance the weapons programs. UNSCR 2371 (2017) prohibits member states from increasing their use of DPRK exported labor. UNSCR 2375 (2017) prohibits entering into new contracts for DPRK exported labor (existing contract completion is allowed). Introduces a ban on Member States from providing work authorizations for DPRK nationals, other than those for which written contracts have been finalized prior to September 11, 2017. UNSCR 2397 (2017) requires member states to repatriate DPRK exported labor within 2 years (with exceptions).	Sec. 104(b)(1)(L), NKSPEA, as amended, authorizes (but does not require) the President to designate any person who facilitates the export of DPRK labor. Sec. 302A, NKSPEA, as amended, blocks import into the U.S. of any goods made by DPRK labor (in violation of Sec. 307, Tariff Act of 1930; 19 U.S.C. 1307, relating to use of prison or coerced labor). Sec. 302B, NKSPEA, as amended, requires the President to impose sanctions on any foreign person who employs DPRK labor.	See EO 13570 (2011), above. See EO 13722 (2016), above.
Trade Ban On Luxury Goods	UNSCR 1718 (2006) prohibits the sale, supply, or transfer of "luxury goods" (but does not define the term). UNSCR 2094 (2013) establishes the first luxury good list. UNSCR 2270 (2016) expands the definition of "luxury goods."	Sec. 104(a)(3), NKSPEA, requires President to designate any person who trades in luxury goods to/from DPRK.	On January 26, 2007, the Department of Commerce issued foreign policy controls for DPRK that included luxury goods. See 72 FR 3722-3730.

Appendix B. (Continued)

	U.N. Security Council Resolution Requirements	U.S. Law	Executive Order
Trade Ban On Luxury Goods	UNSCR 2321 (2016) expands the above list of prohibited goods.		EO 13551 (2010) blocks the assets of specific DPRK individuals and entities, pursuant to emergency authorities and requirements of UNSCRs 1718 (2006) and 1874 (2009). Also blocks assets of anyone found to: --engage in arms trade with DPRK; --provide training, advice, or other services related to arms manufacture, maintenance, or use; --trade in luxury goods; --engage in money laundering, counterfeiting of goods or currency, trafficked in bulk cash, narcotics or other illicit economic activity; or --engage in sanctions evasion. EO 13570 (2011) prohibits importation into the United States from DPRK of any goods, services, or technology. EO 13810 of September 20, 2017, authorizes sanctions on anyone who engages in "at least one significant importation from or exportation to North Korea...."

	U.N. Security Council Resolution Requirements	U.S. Law	Executive Order
Other Trade And Investment	UNSCR 2270 (2016) expands the list of prohibited goods—initially identifying arms, dual-use goods, and luxury goods (2006)— to now include "any item, except food or medicine, if the State determines that such item could directly or indirectly contribute to the development of the DPRK's operational capabilities of its armed forces...." UNSCR 2270 (2016) prohibits participating in joint ventures with designated entities or individuals. UNSCR 2375 (2017) expands the prohibition on joint ventures and cooperative entities to apply to all DPRK entities or individuals, except China-DPRK hydroelectric power infrastructure projects and the Russia-DPRK Rajin-Khasan port and rail project, used for transshipment of Russia-origin coal.		See EO 13570 (2011), above. See EO 13722 (2016), above. EO 13810 of September 20, 2017, authorizes sanctions on anyone who engages in "at least one significant importation from or exportation to North Korea...."
Other Targets		Sec. 104(b)(1), NKSPEA, authorizes (but does not require) the President to designate any person who facilitates bribery or corruption by the DPRK government.	

Appendix B. (Continued)

	U.N. Security Council Resolution Requirements	U.S. Law	Executive Order
Other Targets	UNSCR 2375 (2017) prohibits the export by the DPRK of textiles (including fabrics and partially or fully completed apparel products).	Sec. 104(b)(1)(E), NKSPEA, as amended, authorizes (but does not require) the President to designate any person who trades in textiles with DPRK.	
	UNSCR 2375 (2017) requires a full ban on the supply, sale, or transfer of all condensates and natural gas liquids to DPRK. UNSCR 2375 (2017) limits all refined petroleum products in terms of the amount allowed (for supply, sale, or transfer to DPRK) with very specific preconditions and follow-up action required by Member States, the 1718 Committee and the Committee Secretary. UNSCR 2375 (2017) imposes 12-month caps, based on the previous year's transactions to form a baseline, on the supply, sale, or transfer of crude oil to DPRK. UNSCR 2397 (2017) prohibits providing crude oil above livelihood limits. UNSCR 2397 (2017) prohibits providing refined petroleum products, including diesel and kerosene, above livelihood limits determined by previous trade.	Sec. 104(b)(1)(H), NKSPEA, as amended, authorizes (but does not require) the President to designate any person who facilitates transfers of crude oil, petroleum, liquefied natural gas, other natural gas to the DPRK government.	

	U.N. Security Council Resolution Requirements	U.S. Law	Executive Order
Other Targets		Sec. 104(b)(1)(I), NKSPEA, as amended, authorizes (but does not require) the President to designate any person who engages in DPRK government- sponsored online commerce, including online gambling. Sec. 104(b)(1)(J), NKSPEA, as amended, authorizes (but does not require) the President to designate any person who purchases fishing rights from the DPRK government.	EO 13810 of September 20, 2017, authorizes sanctions on anyone who operates in construction, energy, financial services, fishing, information technology, manufacturing, medical, mining, textiles, or transportation industries in the DPRK.

In: Key Congressional Reports for June 2019 ISBN: 978-1-53616-666-8
Editor: Tara T. Robertson

Chapter 4

POLAND: BACKGROUND AND U.S. RELATIONS*

Derek E. Mix

ABSTRACT

Over the past 30 years, the relationship between the United States and Poland has been close and cooperative. The United States strongly supported Poland's accession to the North Atlantic Treaty Organization (NATO) in 1999 and backed its entry into the European Union (EU) in 2004. Poland has made significant contributions to U.S.- and NATO-led military operations in Iraq and Afghanistan, and Poland and the United States continue to work together closely on a range of foreign policy and international security issues.

Domestic Political and Economic Issues

The 2015 Polish parliamentary election resulted in a victory for the conservative-nationalist Law and Justice party (PiS), which won an

* This is an edited, reformatted and augmented version of Congressional Research Service Publication No. R45784, dated June 25, 2019.

absolute majority of seats in the lower house of parliament (*Sejm*). Mateusz Morawiecki (PiS) is Poland's prime minister and head of government. The center-right Civic Platform (PO) party led the government of Poland from 2007 to 2015. Since winning the election, Law and Justice has made changes to the country's judicial system and enacted other reforms that have generated concerns about backsliding on democracy and triggered an EU rule-of-law investigation.

Poland's next parliamentary election is due to occur in October or November 2019. European Parliament and regional election results indicate that support for Law and Justice remains strong, and the party is favored to win the 2019 election.

Law and Justice candidate Andrzej Duda won Poland's 2015 presidential election. The president is Poland's head of state and exercises a number of limited but important functions. The next presidential election is due to occur in May 2020.

Poland was one of the few EU economies to come through the 2008-2009 global economic crisis without major damage. As an EU member Poland is obligated to adopt the euro as its currency, but it has not set a target date for adoption and continues to use the złoty as its national currency.

Defense Modernization

Poland has been implementing an armed forces modernization plan since 2013, and it intends to spend approximately $49 billion on military equipment acquisitions and upgrades over the period 2017-2026. Completed and prospective purchases from U.S. suppliers, including advanced Patriot missiles and F-35 Joint Strike Fighters, have a large role in this initiative. Poland is one of seven NATO members to meet the alliance's benchmark of spending at least 2% of gross domestic product (GDP) on defense, and it plans to reach 2.5% of GDP by 2030.

Defense Cooperation

Under the United States' European Deterrence Initiative (EDI) and the U.S. military's Operation Atlantic Resolve, as well as NATO's Enhanced Forward Presence mission, U.S. forces have expanded their presence in Poland since 2014 and increased joint training and exercises with their Polish counterparts. While U.S. forces participate in these missions on a rotational basis, the Polish government has proposed the establishment of a permanent U.S. base on Polish territory.

Visa Waiver Program

Although relations between Poland and the United States are largely positive, Poland's exclusion from the U.S. Visa Waiver Program (VWP) has been a point of contention for many years. Some Members of Congress have advocated extending the VWP to include Poland.

Relations with Russia

Relations between Poland and Russia have long been tense, and Polish leaders have tended to view Russian intentions with wariness and suspicion. Poland remains a leading advocate for forceful EU sanctions against Russia over its 2014 annexation of Ukraine's Crimea region and fostering of separatist conflict in eastern Ukraine.

Energy Security

Poland has promoted European energy integration, including projects to expand pipeline and electric grid interconnectivity in order to decrease reliance on Russia. Poland is a leading critic of Nord Stream 2, a Russian-owned pipeline project that would allow Germany to increase the amount of natural gas it imports directly from Russia via the Baltic Sea.

Outlook and Issues for Congress

Given its role as a close U.S. ally and partner, Poland and its relations with the United States are of continuing congressional interest. The main areas of interest include defense cooperation, energy security, and concerns about rule-of-law and governance issues.

Introduction and Issues for Congress

Many U.S. officials and Members of Congress consider Poland to be a key ally of the United States and one of most pro-U.S. countries in Europe. According to the U.S. State Department, areas of close bilateral cooperation with Poland include "NATO capabilities, counterterrorism, nonproliferation, missile defense, human rights, economic growth and innovation, energy security, and regional cooperation in Central and Eastern Europe."[1]

Sources: Created by CRS using data from the U.S. Department of State and ESRI. Fact information from International Monetary Fund, World Economic Outlook Database (April 2019) and CIA World Factbook.

Area: Land area is about 120,728 sq. mi.; slightly smaller than New Mexico.

Population: 37.977 million.

Ethnicity: 96.9% Polish.

Languages: Polish is the official language and first language of 98.2% of the population.

Religion: 85.9% Catholic, 12.1% listed as unspecified.

Gross Domestic Product, 2018 (current prices): $586.015 billion; per capita GDP $15,431.

Currency: złoty (PLN), $1=approx. PLN 3.8, €1=approx. PLN 4.25.

Political Leaders: President: Andrzej Duda; Prime Minister: Mateusz Morawiecki; Foreign Minister: Jacek Czaputowicz; Defense Minister: Mariusz Błaszczak.

Figure 1. Poland at a Glance: Map and Basic Facts.

[1] U.S. Department of State, *U.S. Relations With Poland Fact Sheet*, June 10, 2019.

The Congressional Caucus on Poland is a bipartisan group of Members of Congress who seek to maintain and strengthen the U.S.-Poland relationship and engage in issues of mutual interest to both countries.[2]

Of the Central European and Baltic countries that have joined the North Atlantic Treaty Organization (NATO) and the European Union (EU), Poland is by far the most populous, has the largest economy, and is the most significant military actor. In 1999, with strong backing from the United States, Poland was among the first group of post-communist countries to join NATO. In 2004, again with strong support from the United States, it was among a group of eight post-communist countries to join the EU. Many analysts assert that Poland, more than many other European countries, continues to look to the United States for foreign policy leadership.

Recently, developments related to Russia's resurgence and the attendant implications for U.S. policy and NATO are likely to have continuing relevance for Congress. A variety of factors make Poland a central interlocutor and partner for the United States in examining and responding to these challenges. Since Poland's 2015 parliamentary election, some Members of Congress also have expressed concerns about trends in the country's governance, discussed below.

DOMESTIC OVERVIEW

Political Dynamics

The government of Poland is led by Prime Minister Mateusz Morawiecki of the conservative-nationalist Law and Justice party (PiS). Law and Justice won the October 2015 parliamentary election with 37.6% of the vote, giving the party 235 of the 460 seats in the *Sejm* (lower house of

---

[2] For the 116th Congress, the co-chairs of the Congressional Caucus on Poland are Representative Marcy Kaptur, Representative Dan Lipinski, Representative Jackie Walorski, and Representative Chris Smith.

parliament).[3] This was the first time since the end of communist rule in 1989 that a single party secured an absolute majority in parliament. Law and Justice had spent the previous eight years in opposition after leading the government from 2005 to 2007. The center-right Civic Platform (PO) party, which led the government of Poland from 2007 to 2015, came in second place in the 2015 election with 24.1% of the vote, dropping from 207 to 138 seats in the *Sejm*. The next parliamentary election is due to take place in October or November 2019.[4]

Poland's president is Andrzej Duda, who was the Law and Justice-backed candidate in the May 2015 presidential election. Law and Justice gained momentum five months prior to the parliamentary election with Duda's unexpected victory over the Civic Platform-supported incumbent. The president, who serves a five-year term, is Poland's head of state and resigns party membership upon election. The president exercises functions including making formal appointments, overseeing the country's executive authority, influencing legislation, representing the state in international affairs, and acting as commander-in-chief of the armed forces.

Jarosław Kaczyński is head of Law and Justice and a member of the *Sejm*. Despite his holding no formal post in the government, many observers assert that Kaczyński remains the most powerful politician in Poland who, as party chairman, exerts considerable influence behind the scenes.[5] Jarosław Kaczyński co-founded Law and Justice with his twin brother Lech in 2001. Lech Kaczyński was the president of Poland from 2005 to 2010, when he died in an airplane crash in Russia that also killed 95 other people, including many high-ranking Polish officials.

---

[3] Corinne Deloy, The Law and Justice Party Win the Parliamentary Elections and the Absolute Majority, Robert Schuman Foundation, October 25, 2015.

[4] Members of the *Sejm* are elected to four-year terms through an open-list proportional representation system in which there are 41 multi-member constituencies with between 7 and 20 members each. Members of the Polish Senate (upper house) are elected to four-year terms by plurality vote in 100 single-member constituencies (the Law and Justice party currently has 61 seats in the Senate, and the Civic Platform party has 34).

[5] Henry Foy, "Jaroslaw Kaczynski: Poland's Kingmaker," *Financial Times*, February 26, 2016.

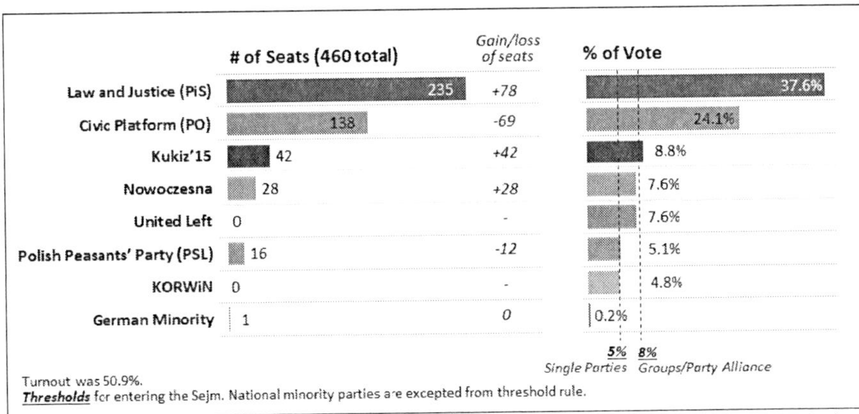

Source: Created by CRS. Information from Poland National Electoral Commission, http://parlament2015.pkw.gov.pl/349_Wyniki_.

Figure 2. Results of the 2015 Polish Parliamentary Election (*Sejm*).

A number of factors contributed to the 2015 election outcome. Law and Justice tapped into public unease over surging non-European migration to Europe by criticizing Civic Platform's willingness to accept migrants under an EU relocation plan. Law and Justice also appeared to gain support by advocating increased public spending for social support programs benefitting families with children, lower-income citizens, and the elderly. During the campaign, the party argued that the benefits of Poland's economic development had fallen unevenly across society and failed to reach many ordinary citizens.

At the same time, observers believe there was a sense of voter fatigue toward Civic Platform and, relatedly, public discontent with the country's political establishment. Civic Platform was damaged by a scandal in which secretly recorded conversations led to the resignation of several government officials in 2015. A changeover in leadership with the 2014 appointment of then-Prime Minister Donald Tusk, who co-founded Civic Platform, as President of the European Council in Brussels was also a factor in the party's decline.

More broadly, the 2015 election and its aftermath appeared to confirm the observation that Polish politics have become characterized by an entrenched social divide between national-oriented social conservatives, represented by Law and Justice, and Western-oriented liberals, represented by Civic Platform.[6]

Since taking office, the Law and Justice-led government has implemented numerous reforms that have proved contentious and raised tensions with the EU as well as domestic opponents; these reforms also have elicited some concern from the United States. Many members of Law and Justice maintain that Poland's post-communist development has been based in part on flawed institutions and values, and Law and Justice leaders interpreted the 2015 election results as a mandate to enact substantial reforms to the country's political system and public institutions. Some argue, therefore, that the party seeks to reduce the influence on national institutions of so-called liberal and secular "European" values and to recast those institutions in ways that promote what the party and its supporters view as traditional national-patriotic values, including close ties with the Catholic Church. Law and Justice also fiercely condemns the communist era and those associated with it, and the party holds a nationalist-oriented worldview that includes enduring suspicion toward Russia and unresolved tensions with Germany.

The results of regional elections in October 2018 and European Parliament (EP) elections in May 2019 indicate that support for the Law and Justice party has held relatively steady since Poland's 2015 election.

- In the 2018 regional elections, Law and Justice won 34% of the vote and the most seats in 9 out of the country's 16 regional assemblies (with an absolute majority in 6). Previously, Law and Justice controlled one regional government.[7] Law and Justice did well among more rural and less affluent voters, while a coalition of opposition parties including Civic Platform did well among more liberal and urban voters. Law and Justice won 4 out of 107

[6] Marc Santora, "Poland Marks an Anniversary, Not in Solidarity, but Division," *The New York Times*, June 4, 2019, and Monika Scislowska, "Poland's Democracy Anniversary Exposes Political Divisions," *The Washington Post*, June 4, 2019.

[7] A coalition of opposition parties including Civic Platform won 27% of the vote and the most seats in Poland's seven other regional assemblies.

municipal elections. The opposition won mayoral races in Poland's largest cities, including Warsaw, Kraków, Wrocław, and Gdańsk.[8]

- In the May 2019 EP elections, Law and Justice came in first place, winning 27 seats with approximately 45% of the Polish vote. A coalition of opposition parties including Civic Platform won 22 seats with approximately 38% of the vote.[9]

Despite numerous public protests over the past three years against the government's reforms, critics observe that opposition parties including Civic Platform have struggled to offer an effective alternate message.[10] Support for Law and Justice, meanwhile, appears to have been mostly unaffected by controversy over its domestic reforms or by a series of corruption scandals reported in late 2018 and early 2019.[11] Given its close association with the Catholic Church, the party came under pressure prior to the EP election with the release of a documentary film about the sexual abuse of children by Polish priests and subsequent efforts to cover up those crimes.[12] After the film was released, the government adopted increased prison sentences for those convicted of sexual abuse of a child.

Controversial Reforms and Tensions with the EU

The most prominent and controversial set of reforms undertaken by the Law and Justice-led government concerns the judicial system. Critics charge that several moves enacted since late 2015 subvert institutional checks and balances, undermine judicial independence and the rule of

---

[8] Joanna Berendt, "Poland Elections Reveal a Deeply Divided Nation," *New York Times*, November 5, 2018.

[9] Aleks Szczerbiak, "Polish Politics after Europe's Election," *Emerging Europe*, June 3, 2019.

[10] Daniel Tilles and Tom Junes, "Poland's Opposition Has Nobody to Blame but Itself," *Foreign Policy*, October 26, 2018.

[11] Michal Broniatowski, "Banking Corruption Scandal Throws Polish Politics into Turmoil," *Politico.eu*, November 13, 2018, Wojciech Kosc, "Central Bank Pay Scandal Hits Poland's Ruling Party," *Politico.eu*, January 9, 2019, and Monika Scislowska, "Poland's Political Leader on Unprecedented Defensive," *AP*, February 16, 2019.

[12] Magdalena Gwozdz-Pallokat, "Documentary on Child Sex Abuse by Catholic Priests Stirs Debate in Poland," *Deutsche Welle*, May 15, 2019.

law, and place the country's courts under political control.[13] The reforms have significantly increased executive and parliamentary powers to select and remove judges, decisions that previously were determined internally by professional bodies within Poland's judiciary. Law and Justice leaders, who blamed the courts for blocking many of the party's legislative priorities when it previously led the government (2005-2007), maintain that the judicial system needed extensive reform because it was slow and inefficient, judges were not properly re-vetted after the transition from communism to democracy, and procedures for selecting new judges lacked fairness and accountability.

Beyond the judicial system, a law adopted in 2016 granted the government the power to hire and fire management of public broadcasting stations, a function previously performed by an independent media supervisory committee. The government maintained that the move was needed to correct political bias and restore balance in the public media. Critics argue that it compromises the independence of state media and relegates it to publicizing the government's official narrative.[14] The government also has cut public funding to some civil society organizations, particularly those supporting migrants and refugees. Critics charged that this move was intended to stifle opponents of government policies.[15]

In 2018, Poland adopted reforms to the country's electoral system. The government asserted that these changes, expected to take effect after the 2019 parliamentary elections, would increase fairness and transparency. Opponents argued that they would politicize the administration of elections and were intended to advantage Law and Justice.[16] The reforms replace seven of the nine members (currently all judges) of the National Electoral

---

[13] For details, see Brittany Benowitz, *Threats to Judicial Independence – Not Discussion of the Holocaust – Are the Real Threat to Polish Democracy*, American Bar Association, January 19, 2019 and Christian Davies, *Hostile Takeover: How Law and Justice Captured Poland's Courts*, Freedom House, May 2018.

[14] Matthew Day, "Poland's President Signs Controversial New Media Law," *The Daily Telegraph*, January 7, 2016.

[15] Claudia Ciobanu and Wojciech Kosc, "Warsaw Grabs Purse Strings of Polish NGOs," *politico.eu*, December 8, 2017.

[16] James Shotter, "Poland's Electoral Reforms Criticised as Return to Communist Times," *Financial Times*, March 25, 2019 and "President Signs Changes to Polish Electoral Law; Opposition Furious," *Radio Poland*, January 16, 2018.

Commission (responsible for conducting and overseeing all elections in Poland) with new members chosen by the *Sejm* according to party proportion. The reforms also call for the National Electoral Commission to appoint new local election commissioners, who are no longer required to be independent of political parties.

Overall, domestic political opponents and outside observers have expressed concern that the actions taken by the government amount to a rollback of Poland's democracy and a program to construct an "illiberal" state.[17] Law and Justice leaders and supporters dispute this portrayal, alleging that their political opponents have crafted this narrative in an attempt to undo the results of the 2015 election and block the government's ability to implement its agenda.

In 2016, the European Commission (the EU's executive institution) launched an inquiry into the effects of the judicial and public media reforms on the rule of law in Poland. The EU subsequently set a series of deadlines for Poland to respond to recommended amendments that would address EU concerns about the ability of the executive and legislature to interfere with the independence of the judiciary. In 2016 and 2017, the Polish government consistently rejected the EU's recommended measures, objecting that the EU was interfering with the country's sovereignty and did not fully understand the Polish legal system. In December 2017, the European Commission recommended the EU move toward imposing an "Article 7" sanction, under which Poland's voting rights in the Council of the EU could be suspended.[18] The measure is unlikely to be enacted, however; Hungary, which has similar Article 7 issues with the EU, has said it would veto the imposition of such a sanction against Poland, which requires unanimity in the Council.

---

[17] See, for example, Melissa Hooper, *Poland* in *The Anatomy of Illiberal States: Assessing and Responding to Democratic Decline in Turkey and Central Europe*, Brookings Institution, February 2019, pp. 16-20, and "Poland's Ruling Law and Justice Party Is Doing Lasting Damage," *The Economist*, April 21, 2018.

[18] European Commission, "Rule of Law: European Commission Acts to Defend Judicial Independence in Poland," press release, December 20, 2017. The Council of the EU (also called the Council of Ministers) is the EU institution representing the national member state governments. See CRS Report RS21372, *The European Union: Questions and Answers*, by Kristin Archick.

The EU also has been developing plans to link the amount of regional funding allocated to Poland (and other countries, such as Hungary) to judicial independence and rule-of-law standards in the next EU budget framework.[19] Poland is the largest beneficiary of funding from the EU budget. In the EU's 2014-2020 budget framework, €106 billion (approximately $120 billion) was allocated to Poland, with the majority of EU support funding regional and municipal infrastructure development.[20]

In October 2018, the Polish government complied with a ruling by the European Court of Justice (ECJ) ordering the suspension of a law that allowed the president to decide whether to retire Supreme Court judges over the age of 65. (The law affected 28 of 72 judges sitting on the appellate panels of the country's Supreme Court at the time it came into effect in July 2018.)[21] The episode marked the first time Law and Justice backtracked on any major element of its controversial reform program. In April 2019, the European Commission launched a new complaint alleging that Poland's process for disciplinary proceedings against judges, enacted in 2017, infringes on EU requirements for judicial independence from political control.[22]

Migration policy has been another source of tension between Poland and the EU. Poland has been a leading opponent of EU policies attempting to relocate migrants and refugees throughout the member states. In 2015, the Civic Platform-led government voted to approve a mandatory EU relocation plan, agreeing to take in approximately 4,600 migrants from outside the EU. The agreement became a significant campaign issue in Poland's 2015 election, with debates about the migration crisis highlighting divisions in Polish society and politics.

---

[19] Lili Bayer, "European Parliament Backs Plan to Link EU Funds to Rule of Law," *Politico.eu*, January 17, 2019.

[20] See Michał Strzałkowski, "European Funds: What Next for the Polish Economy?," *EURACTIV.com*, October 25, 2018, and Aleksandra Stanisławska, *EU Funds in Poland*, Polish Ministry of Foreign Affairs, at https://polska.pl/ economy/investments-projects/eu-funds-poland/.

[21] "Poland Supreme Court Judges Return to Work after EU Court Ruling," *EURACTIV.com*, October 23, 2018.

[22] European Commission, "Rule of Law: European Commission Launches Infringement Procedure to Protect Judges in Poland from Political Control," press release, April 3, 2019.

Law and Justice strongly criticized approval of the plan, and after the terrorist attacks in Paris in November 2015, the incoming Law and Justice-led government indicated that respecting the EU plan was not politically possible. Poland subsequently joined Hungary and the Czech Republic in defying the EU by refusing to the implement the plan, arguing that it infringed on their national sovereignty and that immigration policy was not a competence of the EU. In December 2017, the European Commission referred the three countries to the ECJ over their failure to implement the relocation plan.

Despite these tensions, Jarosław Kaczyński has stated that Law and Justice does not intend to take Poland out of the EU. Surveys show that a large majority of the Polish public views EU membership as beneficial.[23]

The Economy

Poland's economy is among the most successful in Central Europe. Starting with post-communist reform programs in the 1990s and continuing beyond Poland's accession to the EU in 2004, pro-market policies and stable institutions have underpinned strong economic growth, an expanding private sector, and a steady increase in per capita gross domestic product (GDP).[24] Poland's economy was hurt by the 2008 global financial crisis and the ensuing Eurozone crisis but was less affected than most other EU members. The Polish economy was the only European economy to sustain growth in 2008-2009, and Poland avoided a domestic banking crisis.

---

[23] Andrew Rettman, "Kaczynski: No Question of Polish EU Exit," *euobserver.com*, October 18, 2018.

[24] For background on Poland's economic development since communism, see World Bank, Lessons from Poland, Insights for Poland: A Sustainable and Inclusive Transition to High-Income Status, 2017, and Marcin Piatkowski, How Poland Became Europe's Growth Champion: Insights from the Successful Post-Socialist Transition, Brookings Institution, February 11, 2015.

Although Poland joined the EU in 2004, it is not a member of the Eurozone.[25] Poland continues to use the złoty (PLN) as its national currency, and the Eurozone debt crisis that began in Greece in 2009 dampened Polish enthusiasm for adopting the euro. Under the terms of its EU accession treaty, Poland is bound to adopt the euro as its currency eventually, but there is no fixed target date for doing so.

Economic growth in Poland remains high compared to most other EU members. According to the International Monetary Fund (IMF), growth averaged 3.75% per year over the period 2014-2017 and reached 5.1% in 2018.[26] Unemployment is low, decreasing from 10.3% in 2013 to an expected 3.6% in 2019. Forecasts project growth of 3.8% in 2019 and an average of 2.9% annually over the period 2020-2023.

The main drivers of the Polish economy recently have consisted of strong private consumption, investment derived from EU funding, and increased demand for exports. (Nearly 80% of Poland's exports are to other EU countries, with more than a quarter to Germany.[27]) Near-term risks to growth include a potential reduction in EU funding in the next EU budget framework (2021- 2027) and a broader economic slowdown in the EU that could decrease demand for Polish exports.

After the Civic Platform-led government of 2011-2015 sought to consolidate public finances through tax increases and entitlement cuts, the Law and Justice-led government has taken steps to loosen fiscal policy in order to benefit lower-income households and families, encourage higher birth rates, and appeal to older voters.[28] Under the "Family 500+" program, families are eligible to receive a tax-free monthly subsidy of PLN 500 (approximately $132) per month for their second child and every

---

[25] Nineteen EU member countries that use the euro as their common currency are collectively referred to as the Eurozone.

[26] Unless otherwise cited, economic statistics in this section are from International Monetary Fund, World Economic Outlook Database, April 2019. Figures are rounded to the nearest tenth of a percent.

[27] Krzysztof Bień, *The Best Polish Exports' Goods*, Central European Financial Observer, October 25, 2017. Top categories of Poland's exports in include automobile parts and accessories, products made from synthetic materials, furniture, television and radio equipment, automobiles and engines, and computers.

[28] Remi Adekoya, "Why Poland's Law and Justice Party Remains So Popular," *Foreign Affairs*, November 3, 2017.

subsequent child, with lower-income families eligible starting with their first child.[29] Additionally, the government reversed its predecessor's reform raising the retirement age to 67, returning it to 65 for men and 60 for women. Similar to EU-wide averages, the median age in Poland was approximately 38 years old in 2012 and is expected to be 51 years old in 2050.[30] Declining birth rates and net emigration have been the main factors in demographic change in Poland. The aging of the country's population is expected to have challenging implications for Poland's health care and retirement systems.[31]

Concerns that increased government spending on child support and pensions (as well as on planned increases to defense spending) could negatively affect Poland's public finances have largely been balanced by the country's strong economic growth. The budget deficit was 0.6% of GDP in 2018 and is expected to be 2.2% of GDP in 2019. Public debt was approximately 43.6% of GDP in 2018, according to the IMF. (EU rules stipulate that deficits remain below 3% of GDP and that debt remain below 60% of GDP).

Defense Modernization

Poland has repeatedly been invaded by external powers throughout its history. These experiences continue to shape Poland's security perceptions. Territorial defense is the core mission of the Polish military, and Poland's current security strategy is focused primarily on deterring potential Russian aggression. Armed forces modernization, NATO membership, and close ties with the United States are the main components of this strategy. Poland has sought to build a multilayered security policy around this foundation, with participation in EU defense initiatives and cooperation

---

[29] European Commission, Directorate-General for Employment, Social Affairs & Inclusion, *First Results of Poland's Family 500+ Programme Released*, May 16, 2018.

[30] Xavier Devictor, *Poland: Aging and the Economy*, The World Bank, June 14, 2012.

[31] Magdalena Leszko, PhD, Ludmila Zając-Lamparska, PhD, and Janusz Trempala, PhD, "Aging in Poland," *The Gerontologist*, vol. 55, no. 5 (October 2015).

with regional partners such as the Nordic and Baltic countries, the Visegrád Group, and the Bucharest Nine.[32]

Poland has the ninth-largest army in NATO, with 61,200 active personnel. In all, Poland has 117,800 total active military personnel across all branches of the armed forces.[33] Poland ended military conscription in 2009. Poland is one of seven NATO countries meeting the alliance's recommendation of allocating 2% of GDP for defense spending. According to NATO, Polish defense expenditures were 2.05% of GDP ($12.156 billion) in 2018.[34] The Polish government plans to raise defense spending to 2.1% of GDP in 2020 and to gradually increase defense spending to 2.5% of GDP by 2030.[35]

In 2016, the Polish Defense Ministry announced a revised "Technical Modernization Plan" prioritizing air defense, navy, cybersecurity, tanks and armored vehicles, and territorial defense capabilities. From 2017 to 2022, the plan called for approximately $14.5 billion in spending on weapons and equipment acquisition, including new air defense systems, helicopters, UAVs, coastal defense vessels, minesweeper ships, and submarines.[36] In February 2019, the defense ministry announced that it had revised and expanded the plan to include approximately $49 billion in spending on armed forces modernization over the period of 2017-2026. Priorities in the revised plan include short-range anti-aircraft missiles, attack helicopters, submarines, cybersecurity, and the acquisition of fifth-generation combat aircraft.[37]

While foreign purchases continue to play a large role, the Polish government has linked the defense modernization program with efforts to

---

[32] The Visegrád Group (or V4) is a platform for regional cooperation between Poland, the Czech Republic, Hungary, and Slovakia. The Bucharest Nine refers to the countries on NATO's eastern flank (Bulgaria, Czech Republic, Estonia, Hungary, Latvia, Lithuania, Poland, Romania, and Slovakia).

[33] International Institute for Strategic Studies, *The Military Balance 2019*.

[34] NATO Public Diplomacy Division, *Defence Expenditures of NATO Countries (2011-2018)*, March 14, 2019.

[35] "Polish President Signs Defence Spending Boost into Law," *Radio Poland*, October 23, 2017.

[36] Jaroslaw Adamowski, "Polish Defence Ministry Unveils $14.5B Modernization Program," DefenseNews, December 5, 2016.

[37] "Technical Modernization Plan - A Road Map for the Development of the Polish Army," *Defense-Aerospace.com*, February 28, 2019.

develop Poland's defense-industrial base, seeking contracts and partnerships that include local manufacturing and technology transfers.

Another initiative of the Law and Justice-led government has been the establishment of a new territorial defense force, intended to eventually consist of 53,000 volunteers trained and equipped for tasks such as critical infrastructure protection and unconventional warfare.[38]

RELATIONS WITH THE UNITED STATES

Since the end of the Cold War, Poland and the United States have had close relations. The United States strongly supported Poland's accession to NATO in 1999. Warsaw has been an ally in global counterterrorism efforts and contributed large deployments of troops to both the U.S.-led coalition in Iraq and the NATO-led mission in Afghanistan. Links between the United States and Poland are further anchored by extensive cultural ties; approximately 9.6 million Americans are of Polish heritage.

The Law and Justice-led government has sought to cultivate ties with the Trump Administration. In a visit to the United States in September 2018, Polish President Duda suggested that a permanent U.S. military base in Poland might be named "Fort Trump."[39] On February 13-14, 2019, Poland and the United States co-hosted the "Ministerial to Promote a Future of Peace and Security in the Middle East," a conference attended by Vice President Mike Pence and Secretary of State Michael Pompeo.[40]

President Trump earlier delivered a speech in Warsaw on July 6, 2017. The president's remarks on NATO, Russia, U.S.-Polish ties, and Poland's

---

[38] Justyna Pawlak and Kacper Pempel, "In Training with Poland's Volunteer Militia," Reuters, October 18, 2018.

[39] Alan Cowell, "Fort Trump? Poland Makes a Play for a U.S. Military Base," *The New York Times*, September 19, 2018.

[40] See U.S. Department of State, *Ministerial to Promote a Future of Peace and Security in the Middle East* and The White House, *Remarks By Vice President Pence at the Warsaw Ministerial Luncheon,* February 14, 2019. Also see Jacek Czaputowicz and Michael R. Pompeo, "Pompeo and Foreign Minister of Poland: A New Solidarity," *CNN*, February 12, 2019, and Patrick Wintour, "European Powers to Present Cool Front at Warsaw Summit," *The Guardian*, February 12, 2019.

resilience throughout history were well received by many Polish observers, and especially by the Polish government and its supporters. At the same time, critics asserted that the tone of the President's visit, during which he apparently did not raise concerns about Poland's domestic policies, emboldened the government to move ahead with controversial new judicial bills shortly afterward.[41]

While relations between Poland and the United States remain largely positive, there have been points of tension over the past several years. Following President Trump's Warsaw speech, the U.S. State Department released a statement expressing concern about judicial independence and the rule of law in Poland.[42] Some Members of Congress also have expressed concerns about the Polish government's judicial and media reforms. In February 2016, for example, Senators McCain, Durbin, and Cardin co-authored a letter urging Poland to "recommit to the core principles of the [Organization for Security and Cooperation in Europe] and the EU, including the respect for democracy, human rights, and rule of law."[43]

U.S. officials (along with many of their European and Israeli counterparts) objected to controversial Holocaust-related legislation (amendment to the Act on the Institute of National Remembrance) passed by Poland's parliament and signed by President Duda in early 2018.[44] The legislation initially criminalized attributing responsibility for Nazi crimes to the Polish state or nation, potentially punishable by a prison sentence of up to three years, with exemptions for art and academic research. Under

---

[41] See, for example, Slawomir Sierakowski, "Poland Turns Away From Democracy, Thanks to the U.S.," *The New York Times*, July 24, 2017.

[42] The statement asserted that, "The Polish government has continued to pursue legislation that appears to undermine judicial independence and weaken the rule of law in Poland. We urge all sides to ensure that any judicial reform does not violate Poland's constitution or international legal obligations and respects the principles of judicial independence and separation of powers." U.S. Department of State, *Poland: Independence of the Judiciary*, July 21, 2017.

[43] "U.S. Senators Urge Poland to Respect Democracy, Rule of Law," Reuters, February 14, 2016.

[44] See U.S. Department of State, *Legislation In Poland Regarding Crimes Committed During The Holocaust*, January 31, 2018 and U.S. Department of State, *Recent Legislation in Poland*, February 6, 2018.

continued international pressure, the Polish government amended the law in June 2018, making violations a civil (rather than criminal) offense.[45]

In recent years, Polish officials have objected to instances in which commentators and press articles have referred to Auschwitz and other Nazi concentration camps on Polish soil as "Polish death camps," preferring such phrasing as "Nazi concentration camp in German-occupied Poland" (President Obama apologized after using the term "Polish death camp" in 2012).[46] Scholars agree that the term "Polish death camp" is inaccurate and misleading and that the Polish state did not collaborate in the Nazi genocide against Jews.[47] At the same time, historical research has documented instances in which some Poles committed atrocities against Jews during and after World War II. Critics fear the 2018 legislation may serve to stifle debate about such issues and whitewash the culpability of individual Poles in such cases.

In November 2018, a leaked letter from U.S. Ambassador Georgette Mosbacher to Prime Minister Morawiecki reportedly angered some Polish officials by raising concerns about media freedom.[48] The Polish government reportedly had contemplated prosecuting the Polish television station TVN, which is owned by U.S. company Discovery Communications, after it aired footage alleging to show a Polish neo-Nazi group celebrating Adolf Hitler's birthday.

Following the murder of Gdańsk Mayor Paweł Adamowicz in January 2019 by a mentally ill assailant, Representative Marcy Kaptur, a co-chair of the Congressional Caucus on Poland, expressed concern about whether Poland's divided political environment could have played a role in motivating the perpetrator.[49] Adamowicz was a well-known liberal critic of the Law and Justice-led government.

---

45 "Poland Holocaust Law: Government U-turn on Jail Threat," *BBC News*, June 27, 2018.

46 Justin Sink, "Obama Apologizes in Writing for 'Polish Death Camp' Verbal Gaffe," *The Hill*, June 1, 2012.

47 See, for example, Cnaan Liphshiz, "It's Complicated: Inaccuracies Plague Both Sides of 'Polish Death Camps' Debate," *Times of Israel*, January 28, 2018.

48 Vanessa Gera, "U.S. Ambassador Angers Poles with Letter to Prime Minister," Associated Press, November 27, 2018.

49 David M. Herszenhorn, "US Lawmaker Calls for Thorough Probe into Murder of Polish Mayor," *Politico.eu*, January 15, 2019.

In February 2019, Representative Kaptur introduced the Paweł Adamowicz Democratic Leadership Exchange Act of 2019 (H.R. 1270), a bill that would reauthorize the United States-Poland Parliamentary Exchange Program.

Defense Relations

Defense cooperation between Poland and the United States is especially close and extensive. Poland has been a focus of U.S. and NATO efforts to deter potential Russian aggression in the region. In the wake of Russia's aggression against Ukraine starting in 2014, Polish officials reemphasized their wish to permanently base U.S. forces on their territory, despite concerns by some U.S. and European officials that doing so could violate the 1997 NATO-Russia Founding Act.[50] In May 2018, the Polish government released a proposal under which it would contribute $2 billion toward establishing such a base.[51] In a House Armed Services Committee hearing on March 13, 2019, acting Assistant Secretary of Defense for International Security Affairs Kathryn Wheelbarger stated that the related negotiations with Poland were under way.[52] Section 1280 of the John S. McCain National Defense Authorization Act for Fiscal Year 2019 (P.L. 115-232) required the Secretary of Defense to report to the congressional defense committees on the "feasibility and advisability" of permanently stationing U.S. forces in Poland by March 2019. (With discussions between Poland and the U.S. Administration still in progress, the report had not been received as of June 2019.) As part of the United States'

[50] See NATO, Founding Act on Mutual Relations, Cooperation and Security between NATO and the Russian Federation signed in Paris, France, May 27, 1997. The document states that, "NATO reiterates that in the current and foreseeable security environment, the Alliance will carry out its collective defence and other missions by ensuring the necessary interoperability, integration, and capability for reinforcement rather than by additional permanent stationing of substantial combat forces." Proponents of extended permanent basing assert that the "current security environment" with regard to Russia has changed considerably since 1997.

[51] Polish Ministry of National Defence, Proposal for a Permanent U.S. Presence in Poland, 2018.

[52] Kyle Rempfer, "Prepare to Man Fort Trump? U.S. Has Made Poland a 'Very Serious Robust Offer' for Base," MilitaryTimes, March 13, 2019.

missile defense for Europe, an "Aegis-Ashore" site with radar and 24 SM-3 missiles is to become operational in Poland in 2020. Russian officials have characterized the establishment of U.S. missile defense installations in Europe as a "direct threat to global and regional security."[53]

Under the European Deterrence Initiative (EDI), launched in 2014 (originally called the European Reassurance Initiative), the United States has bolstered security in Central and Eastern Europe with an increased rotational military presence, additional exercises and training with allies and partners, improved infrastructure to allow greater responsiveness, enhanced prepositioning of U.S. equipment, and intensified efforts to build partner capacity for newer NATO members and other partners. Approximately 6,000 U.S. military personnel are involved in the associated *Atlantic Resolve* mission at any given time, with units typically operating in the region under a rotational nine-month deployment.[54]

The United States has not increased its permanent troop presence in Europe (currently about 67,000 troops, including two U.S. Army Brigade Combat Teams, or BCTs), but it has rotated additional forces into the region, including nine-month deployments of a third BCT based in the United States.[55] The BCT is based largely in Poland, with units also conducting training and exercises in the Baltic states, Bulgaria, Hungary, and Romania. A combat aviation brigade supports the activities of the BCT.[56] The 4th Infantry Division Mission Command Element, based in Poznań, Poland, acts as the headquarters overseeing rotational units.

Following a meeting between President Trump and President Duda in Washington, DC, on June 12, 2019, President Trump announced that an additional 1,000 troops would be added to the rotational U.S. deployments

---

[53] Andrew E. Kramer, "Russia Calls New U.S. Missile Defense System a 'Direct Threat'," *The New York Times*, May 12, 2016.

[54] See Operation Atlantic Resolve, http://www.eur.army.mil/atlanticresolve/.

[55] The U.S. Army's 1st Armored Brigade Combat Team, 1st Infantry Division arrived in the region from Fort Riley, Kansas in January 2019 with approximately 3,500 personnel, 500 tracked vehicles, and 1,500 wheeled vehicles. See U.S. Army Europe Public Affairs Office, *Fact Sheet: Atlantic Resolve Armored Rotation*, January 11, 2019.

[56] The U.S. Army's 1st Combat Aviation Brigade, 1st Infantry Division arrived in the region from Fort Riley, Kansas in February 2019 with approximately 1,900 personnel, 80 helicopters, and 1,500 pieces of equipment. See U.S. Army Europe Public Affairs Office, *Fact Sheet: Atlantic Resolve Aviation Rotation*, January 22, 2019.

in Poland. The additional troops are expected to come from units based in Germany.[57] The two sides also announced plans for the U.S. military to expand its logistical, administrative, and training infrastructure in Poland, boost the presence of special operations forces, and establish a squadron of aerial reconnaissance drones.[58]

At the 2016 NATO Summit in Warsaw, the alliance agreed to deploy multinational battle groups (approximately 1,100 troops each) to Poland and the three Baltic countries. These "enhanced forward presence" units are intended to deter Russian aggression by acting as a "tripwire" that ensures a response from the entire alliance in the event of a Russian attack. The United States is leading the multinational battalion based in Orzysz, Poland.[59] NATO continues to resist calls to deploy troops permanently in countries that joined after the collapse of the Soviet Union. Accordingly, the enhanced NATO presence has been referred to as "continuous" but rotational.

In recent years, Poland has made a number of significant defense purchases from the United States, and numerous elements of Poland's military equipment modernization plans are of interest and relevance to U.S. defense planners and the U.S defense industry:

- At a February 2019 press conference unveiling the updated Technical Modernization Plan for the Polish armed forces, Polish Defense Minister Mariusz Błaszczak indicated that the procurement of fifth-generation fighter aircraft was a top priority.[60] In May 2019, Poland send a formal letter of request to the United

[57] "Trump: US to Send 1,000 Troops to Poland in New Deal," *BBC News*, June 12, 2019.
[58] C. Todd Lopez, *U.S., Polish Leaders Agree to Increased American Presence in Poland*, U.S. Department of Defense, June 12, 2019.
[59] See North Atlantic Treaty Organization, *Boosting NATO's Presence in the East and Southeast*, January 21, 2019. A U.S. Army armored cavalry squadron accounts for nearly 900 of the approximately 1,200 troops in the Poland-based battlegroup (other contributing countries are Croatia, Romania, and the United Kingdom).
[60] Jacek Siminski, "Poland Making Steps Towards Procuring 32 F-35 Fifth Generation Jets?," *The Aviationist*, February 28, 2019.

States for the purchase of 32 F-35 Joint Strike Fighters, made by Lockheed Martin.[61]

- In February 2019, Poland announced plans to sign a $414 million contract for the purchase of 20 High Mobility Artillery Rocket System (HIMARS) launchers, produced by Lockheed Martin.[62] Delivery is expected by 2023.
- In March 2018, Poland signed a $4.75 billion deal for the purchase of two batteries (four total fire units) of the Patriot integrated air and missile defense system, made by Raytheon.[63] Delivery is expected in 2022.
- In December 2017, the U.S. State Department approved the sale to Poland of F16 support and sustainment services worth up to $200 million, potentially supplied by a number of U.S. contractors.[64]
- In November 2017, the U.S. State Department approved the sale to Poland of up to 150 AIM-120C-7 Advanced Medium Range Air-to-Air Missiles (AMRAAM), made by Raytheon, for an estimated cost of $250 million.[65]
- In November 2016, the U.S. State Department approved the sale to Poland of 70 AGM-158B extended range Joint Air-to Surface Standoff Missiles (JASSM-ER), an air-launched cruise made by Lockheed Martin with a range of approximately 900 kilometers. The deal was worth up to $200 million.[66]
- In 2014, Poland purchased 40 AGM-158A JASSMs (also made by Lockheed Martin) and associated Mid-Life Update (MLU) packages for its F-16C/Ds, reportedly worth about $250 million in

---

[61] Jarosław Adamowski, "Poland Sends Formal Request to Buy F-35s," *DefenseNews*, May 29, 2019.

[62] Jarosław Adamowski, "Poland to Sign $414 Million Deal for Rocket Launchers," *DefenseNews*, February 11, 2019.

[63] Lidia Kelly, "Poland Signs $4.75 Billion Deal for U.S. Patriot Missile System Facing Russia," Reuters, March 28, 2018.

[64] Defense Security Cooperation Agency, *Poland – F-16 Follow-on Support*, News Release, Transmittal No. 17-68, December 19, 2017.

[65] Defense Security Cooperation Agency, *Poland – AIM-120C-7 Advanced Medium Range Air-to-Air Missiles (AMRAAM)*, News Release, Transmittal No. 17-55, November 28, 2017.

[66] Jakub Palowski, "70 JASSM-ER Missiles for the Polish F-16 Fighters. US State Department Issues A Consent," *Defence24.com*, November 29, 2016.

total.[67] The A-variant JASSMs have a range of approximately 370 kilometers.

Economic Ties

Trade between the United States and Poland has increased significantly over the past 15 years. In 2004, for example, U.S. goods exports to Poland were valued at approximately $929 million and imports from Poland were about $1.8 billion. In 2018, U.S. goods exports to Poland were more than $5.4 billion and imports from Poland were more than $8 billion.[68] Leading categories of U.S. exports to Poland include aircraft, machinery, electrical and medical equipment, and vehicles. U.S. imports from Poland represent a wide range of items, including heavy machinery, chemicals, and agricultural products. In 2017, U.S. services exports to Poland were valued at approximately $3.1 billion and services imports from Poland were approximately $2.25 billion.[69]

In 2017, U.S. foreign direct investment in Poland was approximately $12.6 billion.[70] U.S. affiliates employ nearly 200,000 people in Poland.[71] U.S. companies with significant investment in Poland include JP Morgan Chase, Citigroup, Hewlett Packard, UPS, 3M, IBM, United Technologies, General Electric, and Discovery Communications.[72]

Visa Waiver Program

Many Polish officials and citizens continue to express disappointment that the United States has not made Poland a Visa Waiver Program (VWP)

---

[67] "Contract for JASSM Missiles for Poland Awarded," *Defense24.com*, September 16, 2015.
[68] U.S. Department of Commerce, U.S. Census Bureau, *Trade in Goods with Poland.*
[69] U.S. Department of Commerce, Bureau of Economic Analysis, International Transactions, International Services, and International Investment Position Tables, October 19, 2018.
[70] U.S. Department of Commerce, Bureau of Economic Analysis, Direct Investment and MNE.
[71] Daniel S. Hamilton and Joseph P. Quinlan, The Transatlantic Economy 2019, Center for Transatlantic Relations, p. 141.
[72] American Chamber of Commerce in Poland, American Investments in Poland, April 2018.

country.[73] Current U.S. visa policy requires Poles who wish to travel to the United States to apply for a visa by filling out an application, paying a $160 nonrefundable fee, and completing an interview at a U.S. embassy or consulate. These requirements are waived for citizens of most EU countries, since most of the countries qualify to be included in the VWP.[74] The VWP allows for visa-free travel to the United States for up to 90 days. Under U.S. policy, Poland does not meet the VWP's qualifying criteria because its visitor visa refusal rate (the percentage of applications rejected by U.S. consular officers who cannot overcome the refusal) remains above the 3% limit. The refusal rate for Poland was 3.99% in FY2018 and 5.92% in FY2017.[75] Nonimmigrant visas issued to Polish nationals increased nearly 60% from 2009 to 2018.[76]

Citing Poland's status as a close U.S. ally, some Members of Congress have attempted to change the law governing the VWP to allow Poland to qualify. In the 115th Congress, Representative Mike Quigley introduced a bill (H.R. 2388, Poland Visa Waiver Act of 2017) that would have authorized the Secretary of Homeland Security to designate Poland a VWP country. Some opponents of extending the VWP to include Poland argue that such a step could allow a significant increase in the number of Poles who overstay their visas and remain illegally in the United States (i.e., become unauthorized aliens).[77] Proponents of including Poland argue that such a move would increase U.S. tourism revenue, boost public diplomacy, and strengthen national security by extending the information-sharing elements of the VWP to Poland.[78]

---

73 For more information on the VWP and debates surrounding the VWP, see CRS Report RL32221, Visa Waiver Program, by Jill H. Wilson.

74 There are currently 38 countries participating in the U.S. Visa Waiver Program, including 23 of the 28 member countries of the European Union.

75 U.S. Embassy and Consulate in Poland, Fiscal Year 2018 U.S. Visa Refusal Rate for Polish Nationals is 3.99%, February 4, 2019.

76 U.S. State Department, Nonimmigrant Visas Issued by Nationality (Including Border Crossing Cards): Fiscal Years 2009-2018, Report of the Visa Office 2018.

77 See, for example, Jessica Vaughan, *Senators Vote to Allow Visa Waivers for Poland*, Center for Immigration Studies, June 22, 2015.

78 See, for example, Edwin J. Feulner, *Vouching for the Visa Waiver Program*, Heritage Foundation, October 16, 2014.

Relations with Russia

Historically, Poland has had a difficult relationship with Russia. Poland's view of Russia remains affected by the experience of Soviet invasion during World War II and Soviet domination during the communist era. In more recent years, Polish leaders have consistently expressed warnings about the nature of Vladimir Putin's government in Russia, tending to view Russia as a potential threat to Poland and its neighbors. This perception predates Russia's invasion of Georgia in 2008, but events in Ukraine since 2014 have sharpened Polish concerns about Russia's intentions and have put security at the top of Poland's national agenda.

The Law and Justice-led government has maintained a hard line in its approach to Russia and entrenched Poland's position as one of the EU's most hawkish countries on Russia policy. Poland has been one of the leading advocates for adopting and maintaining robust EU sanctions against Russia in response to its actions in Ukraine, although it has been one of the countries most affected by Russian retaliatory sanctions.

One area of particular relevance to Poland's security is Kaliningrad, a 5,800-square-mile Russian exclave wedged between Poland and Lithuania (see *Figure 1*). Ceded to Russia by Germany following World War II, Kaliningrad is a key strategic territory for Russia, allowing it to project military power into NATO's northern flank. The territory has a heavy Russian military presence, including the Baltic Fleet and two airbases. Russia has repeatedly deployed Iskander short-range nuclear-capable missiles in Kaliningrad, and reports indicated that a 2018 deployment could be permanent.[79] According to NATO officials, Russia is using Kaliningrad to pursue an antiaccess/area denial (A2/AD) strategy that involves layering surface-to-air missiles to potentially block off access to the Baltic states and much of Poland.[80] Kaliningrad's geographic isolation

---

[79] "Russia Deploys Iskander Nuclear-Capable Missiles to Kaliningrad," Reuters, February 5, 2018.

[80] Dominik Jankowski, *Six Ways NATO Can Address the Russian Challenge*, Atlantic Council, July 4, 2018.

also allows for a scenario in which Russia tries to seize the Suwałki Gap, the 100-kilometer border between Poland and Lithuania that separates Kaliningrad from Russia's ally Belarus.[81]

ENERGY SECURITY

Poland has been a leading opponent and critic of the Nord Stream 2 pipeline that would allow Germany to increase the amount of natural gas it imports directly from Russia via the Baltic Sea.[82] Poland argues that the completion of Nord Stream 2 would allow Gazprom, Russia's state-owned gas company, to further consolidate its monopoly over the Central European gas market, as Gazprom would have full control of all gas transmission routes and Russian gas would dominate the European network hubs in Germany and Austria. Poland maintains that Russia would further gain geopolitical leverage because it could arbitrarily shift supply corridors in Europe, giving it the ability to continue supplying European markets through Germany while restricting or completely halting gas transit through Poland and/or Ukraine. Polish officials have expressed the view that U.S. involvement, including the adoption of sanctions, is crucial for efforts to oppose construction of the pipeline.[83]

While approximately two-thirds of the natural gas and most of the oil consumed in Poland comes from Russia, Poland continues to rely on domestically produced coal for much of its electricity generation. Russian gas accounts for less than 10% of Poland's primary energy supply.[84] Successive Polish governments have prioritized efforts to diversify the country's energy sources. Poland has been taking steps to expand pipeline

[81] LTG (Ret.) Ben Hodges, Janusz Bugajski, and Peter Doran, *Securing the Suwalki Corridor*, Center for European Policy Analysis, July 2018.

[82] See CRS In Focus IF11138, *Nord Stream 2: A Fait Accompli?*, by Paul Belkin et al.

[83] Nick Snow, "US Sanctions Necessary to Stop Nord Stream 2 Pipeline, Speakers Say," *Oil & Gas Journal*, March 15, 2019.

[84] Statistics from BP Statistical Review of World Energy – all data, 1965-2017, at https://www.bp.com/en/global/corporate/energy-economics/statistical-review-of-world-energy.html.

interconnectivity with its neighbors, including plans to develop the Baltic Pipe project, expected to be operational in 2022, which would connect Poland's gas infrastructure via Denmark to Norwegian supplies.[85] Poland's supply contract with Gazprom expires in 2022, and Poland is unlikely to seek its renewal. Poland also has developed the ability to reverse the flow of gas in the Polish section of the Yamal pipeline, which runs from Russia to Germany via Belarus and Poland, in order to import natural gas from the west in the case of a crisis involving a cut-off of Russian gas from the east. A liquefied natural gas (LNG) terminal on the Baltic Sea coast near the German border (Świnoujście) became operational in 2015, and in October 2018 the Polish state energy company signed a 20-year contract to purchase LNG from a U.S. supplier.[86]

The Polish government also has been a leading advocate for a stronger EU energy policy that reduces collective dependence on Russia. Poland has been active in projects that enhance regional energy security by interconnecting national gas networks through the construction of new pipelines. The construction of new connectors with Slovakia and the Czech Republic is underway, and the planned Gas Interconnection Poland-Lithuania (GIPL), expected to become operational by 2022, would link the natural gas grid of the Baltic countries to the rest of the EU.[87]

Many U.S. officials and Members of Congress have regarded European energy security as a U.S. interest. In particular, there has been concern in the United States over the influence that Russian energy dominance could have on the ability to present a united transatlantic position when it comes to other issues related to Russia. Successive U.S. administrations have encouraged EU member states to reduce energy dependence on Russia through diversification of supply and supported European steps to develop alternative sources and increase energy efficiency.

In the 116th Congress, related bills include the European Energy Security and Diversification Act 2019 (House-passed H.R. 1616 and S. 704) and the Protect European Energy Security Act (H.R. 1081).

---

[85] See Baltic Pipe Project, https://www.baltic-pipe.eu/the-project/.

[86] Radio Free Europe/Radio Liberty, "Poland Seals Deal To Buy LNG From U.S. To Ease Dependence On Moscow," October 17, 2018.

[87] Jo Harper, "Polish Gas Terminal Speaks Volumes," *Forbes*, March 31, 2019.

Introduced by Representative Adam Kinzinger and Senator Christopher Murphy, the European Energy Security and Diversification Act 2019 aims to prioritize and enhance U.S. efforts to encourage European countries to diversify energy sources and supply routes and increase regional energy security. Introduced by Representative Denny Heck, the Protect European Energy Security Act would require reports to Congress by the Secretary of State, Secretary of the Treasury, and the Director of National Intelligence detailing U.S. diplomatic efforts to oppose the construction of Nord Stream 2 and to promote European energy security and decrease European dependence on Russian energy.

CONCLUSION

Poland appears likely to remain a strong U.S. ally and an increasingly important U.S. security partner in Europe. Many analysts believe that close cooperation between the United States and Poland will continue for the foreseeable future in areas such as efforts to deter potential Russian aggression, the future of NATO, energy security, and economic issues. Statements by Polish leaders suggest that Poland is likely to continue looking to the United States for leadership on foreign policy and security issues.

During the 116th Congress, the issue of establishing a permanent U.S. military base in Poland or increasing the size of the U.S. military presence in Poland may remain of interest to Members of Congress. Contracted and prospective U.S. arms sales to Poland, including major items such as the F-35 and Patriot missile systems, also may be of congressional interest. Some Members may wish to consider Poland's status with regard to the U.S. Visa Waiver Program. Poland is likely to be of continuing importance in the area of European energy security.

Members of Congress also may wish to remain informed about legislative, governance, and ruleof-law issues in Poland, including with regard to the numerous controversial domestic reforms enacted over the past several years. Members of Congress may have an interest in monitoring political developments in relation to the Polish parliamentary election due to occur in October or November 2019.

In: Key Congressional Reports for June 2019 ISBN: 978-1-53616-666-8
Editor: Tara T. Robertson

Chapter 5

VENEZUELA: BACKGROUND AND U.S. RELATIONS (UPDATED)*

Clare Ribando Seelke, Rebecca M. Nelson and Rhoda Margesson

ABSTRACT

Venezuela remains in a deep political and economic crisis under the authoritarian rule of Nicolás Maduro of the United Socialist Party of Venezuela. Maduro, narrowly elected in 2013 after the death of Hugo Chávez (president, 1999-2013), began a second term on January 10, 2019, that most Venezuelans and much of the international community consider illegitimate. Since January, Juan Guaidó, president of Venezuela's democratically elected, opposition-controlled National Assembly, has sought to form an interim government to serve until internationally observed elections can be held. Although the United States and 53 other countries recognize Guaidó as interim president, the military high command, supported by Russia and Cuba, has remained loyal to Maduro. Venezuela is in a political stalemate as conditions in the country deteriorate.

* This is an edited, reformatted and augmented version of Congressional Research Service, Publication No. R44841, dated June 4, 2019.

Venezuela's economy has collapsed. It is plagued by hyperinflation, severe shortages of food and medicine, and electricity blackouts that have worsened an already dire humanitarian crisis. In April 2019, United Nations officials estimated that some 90% of Venezuelans are living in poverty and 7 million need humanitarian assistance. Maduro has blamed U.S. sanctions for these problems, but most observers cite economic mismanagement and corruption under Chávez and Maduro for the current crisis. U.N. agencies estimate that 3.7 million Venezuelans had fled the country as of March 2019, primarily to other Latin American and Caribbean countries.

U.S. Policy

The United States historically had close relations with Venezuela, a major U.S. oil supplier, but relations deteriorated under the Chávez and Maduro governments. The Trump Administration has employed targeted sanctions against Venezuelan officials responsible for human rights violations, undermining democracy, and corruption, as well as on individuals and entities engaged in drug trafficking. Since 2017, the Administration has imposed a series of broader economic sanctions on the Maduro government and state oil company *Petróleos de Venezuela, S. A.* (PdVSA).

On January 23, 2019, the Trump Administration recognized the Guaidó government. It has imposed additional sanctions on Maduro officials, blocked Maduro and his government from benefitting from revenue from PdVSA, and imposed secondary sanctions on entities doing business with Maduro. The Administration has provided some $213 million in humanitarian assistance to countries sheltering Venezuelans and pre-positioned emergency supplies for Venezuelans at the country's borders.

Congressional Action

The 116th Congress likely will fund foreign assistance to Venezuela and neighboring countries sheltering Venezuelans. Congress may consider additional steps to influence the Venezuelan government's behavior in promoting a return to democracy and to relieve the humanitarian crisis. In FY2019, Congress provided $17.5 million for democracy and rule of law programs in Venezuela (P.L. 116-6). The Administration's proposed FY2020 budget asks for $9 million in democracy aid and authority to transfer up to $500 million to support a transition or respond to a crisis in Venezuela. On May 20, 2019, the House Appropriations Committee reported its version of the FY2020 Department of State and Foreign Operations Appropriations Act (H.R. 2389), which would provide $17.5 million in democracy and human rights aid to Venezuela.

On May 22, 2019, the Senate Foreign Relations Committee ordered S. 1025 reported with an amendment to include three House-passed measures: H.R. 854 would authorize expanded humanitarian aid, H.R. 920 would prohibit arms transfers to Maduro, and H.R. 1477 would require an assessment of Russian-Venezuelan security cooperation. S. 1025 would increase humanitarian assistance to Venezuela and countries sheltering Venezuelans and provide support for Venezuela's democratic transition. On May 22, 2019, the House Judiciary Committee ordered reported H.R. 549 to make certain Venezuelans in the United States eligible for Temporary Protected Status. On April 9, 2019, the House Committee on Foreign Affairs ordered reported H.R. 1004 to prohibit the use of U.S. military forces in combat operations in Venezuela.

RECENT DEVELOPMENTS

On May 29, 2019, the government of Norway confirmed that the representatives of the main political actors in Venezuela had returned to Oslo "to move forward in the search for an agreed-upon and constitutional solution for the country" (See "Foreign Involvement in Venezuela's Political Crisis," below).

On May 25, 2019, the U.S. Department of State stated that "the only thing to negotiate with Nicolás Maduro is the conditions of his departure.... [If] the talks in Oslo will focus on that objective ... we hope progress will be possible" (See "U.S. Policy," below).

On May 22, 2019, the Senate Foreign Relations Committee ordered S. 1025, a comprehensive bill to address the crisis in Venezuela, to be reported with an amendment to include three House-passed measures: H.R. 854 would authorize expanded humanitarian aid, H.R. 920 would restrict arms transfers to Maduro, and H.R. 1102 would require an assessment of Russian-Venezuelan security cooperation. (See Appendix A, below).

On May 22, 2019, the House Judiciary Committee ordered reported H.R. 549 to make certain Venezuelans in the United States eligible for Temporary Protected Status. A companion bill (S. 636) has been introduced in the Senate (See Appendix A, below).

On May 19, 2019, the interim government led by Juan Guaidó hired a prominent sovereign debt lawyer, Lee Buchheit, on a pro bono basis to help restructure the country's $150 billion in outstanding debt should it obtain complete control in Venezuela (See "Economic Crisis," below).

On May 16, 2019, an ad hoc board of directors of *Petróleos de Venezuela, S. A.* (PdVSA)— Venezuela's state-owned oil company— appointed by Venezuela's interim government made a $71 million interest payment on PdVSA bonds maturing in 2020. Defaulting on the PdVSA 2020 bonds could have resulted in legal challenges by creditors and the seizure of U.S. refiner Citgo, owned by PdVSA (See "Energy Sector Concerns and U.S. Economic Sanctions," below).

On May 15, 2019, the U.S. Department of Transportation ordered flights between the United States and Venezuela to be suspended due to information the agency obtained from the Department of Homeland Security that "conditions in Venezuela threaten the safety and security of passengers, aircraft, and crew traveling to or from that country" (See "U.S. Policy," below).

On May 8, 2019, Venezuelan intelligence agents arrested the vice president of the National Assembly, Edgar Zambrano, reportedly for his involvement in a failed uprising against authoritarian leader Nicolás Maduro on April 30, 2019 (See "Human Rights," below).

On April 30, 2019, Guaidó and Leopoldo López, a former political prisoner and head of the Popular Will (VP) party who had been released from house arrest by pro-Guaidó military officials, called for a civil-military rebellion against Maduro. Forces loyal to Maduro violently put down pro-Guaidó supporters and attacked journalists. López sought refuge in the Spanish embassy (See "Maduro's Second Term Deemed Illegitimate; Interim Government Formed," below).

On April 10, 2019, United Nations humanitarian experts updated the U.N. Security Council on the humanitarian situation in Venezuela, estimating that some 7 million Venezuelans were in need of humanitarian assistance (See "Humanitarian Situation" below).

INTRODUCTION

Venezuela, long one of the most prosperous countries in South America with the world's largest proven oil reserves, continues to be in the throes of a deep political, economic, and humanitarian crisis. Whereas populist President Hugo Chávez (1999-2013) governed during a period of generally high oil prices, his successor, Nicolás Maduro of the United Socialist Party of Venezuela (PSUV), exacerbated an economic downturn caused by low global oil prices through mismanagement and corruption. According to Freedom House, Venezuela fell from "partly free" under Chávez to "not free" under Maduro.[1] On January 10, 2019, Maduro took office for a second term after winning an election in May 2018 that was deemed illegitimate within Venezuela and by the United States and more than 50 other countries.[2] Maduro has resisted domestic and international pressure to leave office and allow an interim government led by Juan Guaidó, president of the National Assembly, to serve until free and fair elections can be convened.[3]

U.S. relations with Venezuela, a major oil supplier, deteriorated during Chávez's rule, which undermined human rights, the separation of powers, and freedom of expression. U.S. and regional concerns have deepened as the Maduro government has manipulated democratic institutions; cracked down on the opposition, media, and civil society; engaged in drug trafficking and corruption; and refused most humanitarian aid.

---

[1] Freedom House, *Freedom in the World: 2019.*

[2] Reuters, "Venezuela's Maduro Reelected Amid Outcry over Vote," May 20, 2018. The United States, the EU, the Group of Seven, and most Western Hemisphere countries do not recognize the legitimacy of his mandate.

[3] In the absence of an elected president, the Venezuelan Congress's president must become acting president of a transition government until elections can be called. Constitution of the Bolivarian Republic of Venezuela, 1999, as amended by Amendment No. 1 of 15 February 2009, Article 233, translation by the Ministry of Communication and Information of the Bolivarian Republic of Venezuela, 2010.

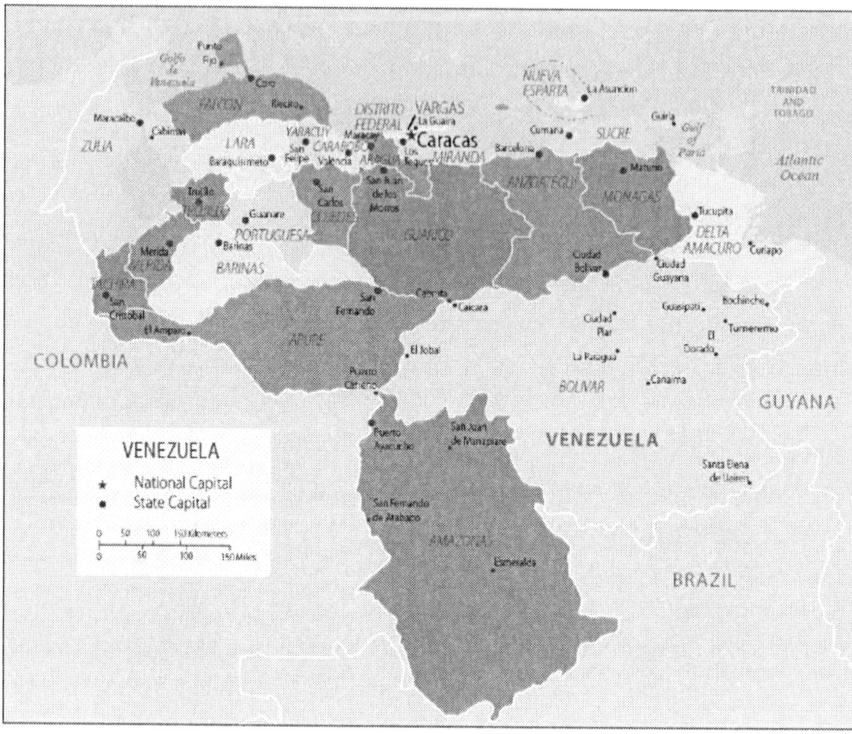

Source: Congressional Research Service (CRS).

Figure 1. Political Map of Venezuela.

Venezuela at a Glance

Population: 28 million (2019 est., IMF)
Area: 912,050 square kilometers (slightly more than twice the size of California)
GDP: $76.5 billion (2019, current prices, IMF est.) GDP Growth: 18% (2018, IMF est.)
GDP Per Capita: $2,724 (2019, current prices, IMF est.)
Key Trading Partners: Exports—U.S.: 35.2%, India: 21.4%, China: 15.2%. Imports—**U.S.:** 51.5%, **China:** 15.7%, Brazil: 13.8% (2019, EIU)
Unemployment: 44.3% (2019, IMF)
Life Expectancy: 74.7 years (2017, UNDP) Literacy: 97.1% (2016, UNDP)
Legislature: National Assembly (unicameral), with 167 members; National Constituent Assembly, with 545 members (United States does not recognize)

Sources: Economist Intelligence Unit (EIU); International Monetary Fund (IMF); United Nations Development Programme (UNDP).

Efforts to hasten a return to democracy in Venezuela have failed thus far, but efforts to find a political solution to the crisis continue.[4]

This chapter provides an overview of the overlapping political, economic, and humanitarian crises in Venezuela, followed by an overview of U.S. policy toward Venezuela.

POLITICAL SITUATION

Legacy of Hugo Chávez: 1999-2013[5]

In December 1998, Hugo Chávez, a leftist populist representing a coalition of small parties, received 56% of the presidential vote (16% more than his closest rival). Chávez's commanding victory illustrated Venezuelans' rejection of the country's two traditional parties, Democratic Action (AD) and the Social Christian party (COPEI), which had dominated Venezuelan politics for the previous 40 years. Most observers attribute Chávez's rise to power to popular disillusionment with politicians who were judged to have squandered the country's oil wealth through poor management and corruption. Chavez's campaign promised constitutional reform; he asserted that the system in place allowed a small elite class to dominate Congress and waste revenues from the state oil company, *Petróleos de Venezuela, S. A.* (PdVSA).

Venezuela had one of the most stable political systems in Latin America from 1958 until 1989. After that period, however, numerous economic and political challenges plagued the country. In 1989, then-President Carlos Andres Pérez (AD) initiated an austerity program that fueled riots in which several hundred people were killed. In 1992, two attempted military coups threatened the Pérez presidency, one led by

---

[4] CRS In Focus IF11216, *Venezuela: International Efforts to Resolve the Political Crisis*, by Clare Ribando Seelke.

[5] This section draws from CRS Report R42989, *Hugo Chávez's Death: Implications for Venezuela and U.S. Relations*, by Mark P. Sullivan; and CRS Report R43239, *Venezuela: Issues for Congress, 2013-2016*, by Mark P. Sullivan.

Chávez, who at the time was a lieutenant colonel railing against corruption and poverty. Chávez served two years in prison for that failed coup attempt. In May 1993, the legislature dismissed Pérez from office for misusing public funds. The election of former President Rafael Caldera (1969-1974) as president in December 1993 brought a measure of political stability, but the government faced a severe banking crisis. A rapid decline in the world price of oil caused a recession beginning in 1998, which contributed to Chávez's landslide election.

Under Chávez, Venezuela adopted a new constitution (ratified by plebiscite in 1999), a new unicameral legislature, and a new name for the country—the Bolivarian Republic of Venezuela, named after the 19th century South American liberator Simón Bolívar. Buoyed by windfall profits from increases in the price of oil, the Chávez government expanded the state's role in the economy by asserting majority state control over foreign investments in the oil sector and nationalizing numerous private enterprises. Chávez's charisma, use of oil revenue to fund domestic social programs and provide subsidized oil to Cuba and other Central American and Caribbean countries, and willingness to oppose the United States captured global attention.[6]

After Chávez's death, his legacy has been debated. President Chávez established an array of social programs and services known as *missions* that helped reduce poverty by some 20% and improve literacy and access to health care.[7] Some maintain that Chávez also empowered the poor by involving them in community councils and workers' cooperatives.[8] Nevertheless, his presidency was "characterized by a dramatic concentration of power and open disregard for basic human rights

---

[6] Chávez viewed himself as a leader of an integrated Latin America struggling against an external power (the United States), similar to how Simón Bolívar led the struggle against Spain in the 19th century. Carlos A. Romero and Víctor M. Mijares, "From Chávez to Maduro: Continuity and Change in Venezuelan Foreign Policy," *Contexto Internacional*, vol. 38, no.1 (2016), pp. 178-188. In 2005, the PetroCaribe program began providing subsidized oil to many Caribbean and Central American countries; however, the volume of shipments declined after 2012. David L. Goldwyn and Cory R. Gill, *The Waning of PetroCaribe? Central American and Caribbean Energy in Transition*, Atlantic Council, 2016.

[7] Daniel Hellinger and Anthony Petros Spanakos, "The Legacy of Hugo Chávez," *Latin American Perspectives*, vol. 44, no. 1, January 2017, pp. 4-15.

[8] Eva Golinger, "Opinion: Chávez was a Maker of Dreams," CNN, March 7, 2013.

guarantees," especially after his brief ouster from power in 2002.[9] Declining oil production, combined with massive debt and high inflation, have shown the costs involved in Chávez's failure to save or invest past oil profits, tendency to take on debt and print money, and decision to fire thousands of PdVSA technocrats after an oil workers' strike in 2002-2003.[10]

Venezuela's 1999 constitution, amended in 2009, centralized power in the presidency and established five branches of government rather than the traditional three branches.[11] Those branches include the presidency, a unicameral National Assembly, a Supreme Court, a National Electoral Council (CNE), and a "Citizen Power" branch (three entities that ensure that public officials at all levels adhere to the rule of law and that can investigate administrative corruption). The president is elected for six-year terms and can be reelected indefinitely; however, he or she also may be made subject to a recall referendum (a process that Chávez submitted to in 2004 and survived but Maduro cancelled in 2016). Chávez exerted influence over all the government branches, particularly after an outgoing legislature dominated by *chavistas* appointed pro-Chávez justices to control the Supreme Court in 2004 (a move that Maduro's allies would repeat in 2015).

In addition to voters having the power to remove a president through a recall referendum process, the National Assembly has the constitutional authority to act as a check on presidential power. The National Assembly consists of a unicameral Chamber of Deputies with 167 seats whose members serve for five years and may be reelected once. With a simple

---

[9] Although President Chávez remained widely popular until mid-2001, his standing eroded afterward amid growing concerns by some that he was imposing a leftist agenda on the country and that his government was ineffective in improving living conditions. In April 2002, massive protests and pressure by the military led to the ousting of Chávez from power for less than three days. He ultimately was restored to power by the military after an interim president alienated the military and the public by taking hardline measures, including the suspension of the constitution. Human Rights Watch, "Venezuela: Chávez's Authoritarian Legacy," March 5, 2013.

[10] Francisco Monaldi, *The Impact of the Decline in Oil Prices on the Economics, Politics, and Oil Industry in Venezuela*, Columbia Center on Global Energy Policy, September 2015, pp. 9-13.

[11] Antonio Ramirez, "An Introduction to Venezuelan Governmental Institutions and Primary Legal Sources," New York University Law School Library, May 2016.

majority, the legislature can approve or reject the budget and the issuing of debt, remove ministers and the vice president from office, overturn enabling laws that give the president decree powers, and appoint the five members of the CNE (for seven-year terms) and the 32 members of the Supreme Court (for one 12-year term).

Nicolás Maduro

A former trade unionist who served in Venezuela's legislature from 1998 until 2006, Nicolás Maduro held the position of National Assembly president from 2005 to 2006, when he was selected by President Chávez to serve as foreign minister. Maduro retained that position until mid-January 2013, concurrently serving as vice president beginning in October 2012, when President Chávez tapped him to serve in that position following his reelection. Maduro often was described as a staunch Chávez loyalist. Maduro's partner since 1992 is well-known Chávez supporter Cilia Flores, who served as the president of the National Assembly from 2006 to 2011; the two were married in July 2013.

With a two-thirds majority, the assembly can remove judges, submit laws directly to a popular referendum, and convene a constitutional assembly to revise the constitution.[12]

Maduro Government: 2013-2018[13]

After Chávez's death in March 2013, Venezuela held presidential elections in April. Acting President Nicolás Maduro defeated Henrique Capriles of the Democratic Unity Roundtable (MUD) of opposition parties by 1.5%. The opposition alleged significant irregularities and protested the outcome.

After his razor-thin victory, Maduro sought to consolidate his authority. Security forces and allied civilian groups violently suppressed protests and restricted freedom of speech and assembly. In 2014, 43 people

---

[12] Ramirez, "An Introduction to Venezuelan Governmental Institutions;" CRS Report R43239, *Venezuela: Issues for Congress, 2013-2016*, by Mark P. Sullivan.

[13] See also CRS Report R43239, *Venezuela: Issues for Congress, 2013-2016*, by Mark P. Sullivan.

died and 800 were injured in clashes between pro-government forces and student-led protesters concerned about rising crime. President Maduro imprisoned opposition figures, including Leopoldo López, head of the Popular Will (VP) party, who was sentenced to more than 13 years in prison for allegedly inciting violence. In February 2015, the government again cracked down.

In the December 2015 legislative elections, the MUD captured a two-thirds majority in Venezuela's National Assembly—a major setback for Maduro. The PSUV-aligned Supreme Court blocked three MUD deputies from taking office, which deprived the opposition of a two-thirds majority. From January 2016 through August 2017 (when the National Constituent Assembly discussed below voted to give itself legislative powers), the Supreme Court blocked laws and assumed the legislature's functions.

In 2016, opposition efforts focused on attempts to recall President Maduro in a national referendum. The government used delaying tactics to slow the process. In October 2016, the CNE suspended the recall effort after five state-level courts issued rulings alleging fraud in a signature collection drive that had amassed millions of signatures. In late 2016, most of the opposition (with the exception of Popular Will) agreed to talks with the government mediated by the Vatican; the former leaders of the Dominican Republic, Spain, and Panama; and the head of the Union of South American Nations. By December 2016, the opposition had left the talks due to a lack of progress on the part of the government in meeting its commitments.[14]

In 2017, the Maduro government continued to harass and arbitrarily detain opponents. In addition, President Maduro appointed a hardline vice president, Tareck el Aissami, former governor of the state of Aragua and designated by the United States as a drug kingpin. Popular protests, which were frequent between 2014 and 2016, had dissipated. In addition to

---

[14] Those commitments included (1) releasing political prisoners, (2) announcing an electoral calendar, (3) respecting the National Assembly's decisions, and (4) addressing humanitarian needs.

restricting freedom of assembly, the government cracked down on media outlets and journalists.[15]

Despite these obstacles, the MUD became reenergized in response to the Supreme Court's March 2017 rulings to dissolve the legislature and assume all legislative functions. After domestic protests, a rebuke by then-Attorney General Luisa Ortega (a Chávez appointee), and an outcry from the international community, President Maduro urged the court to revise those rulings, and it complied. In April 2017, the government banned opposition leader and two-time presidential candidate Henrique Capriles from seeking office for 15 years, which fueled more protests.

From March to July 2017, the opposition conducted large, sustained protests against the government, calling for President Maduro to release political prisoners, respect the separation of powers, and hold an early presidential election. Clashes between security forces (backed by armed civilian militias) and protesters left more than 130 dead and hundreds injured (See "Human Rights" below).

Constituent Assembly Established

In May 2017, President Maduro announced that he would convene a constituent assembly to revise the constitution and scheduled July 30 elections to select delegates to that assembly. The Supreme Court ruled that Maduro could convoke the assembly without first holding a popular referendum (as the constitution requires). The opposition boycotted, arguing that the elections were unconstitutional; a position shared by then-Attorney General Luisa Ortega and international observers (including the United States). In an unofficial plebiscite convened in mid-July by the MUD, 98% of some 7.6 million Venezuelans cast votes rejecting the creation of a constituent assembly; the government ignored that vote. Despite an opposition boycott and protests, the government orchestrated

[15] In February 2017, the government suspended CNN en Español from cable after it aired an investigation into fraudulent Venezuelan passports being sold in Iraq. Associated Press, "Venezuela Shuts Off CNN en Español After Criticizing Channel's Passport-Selling Report," February 15, 2017.

the July 30, 2017, election of a 545- member National Constituent Assembly (ANC) to draft a new constitution.[16]

Many observers viewed the establishment of the ANC as an attempt by the ruling PSUV to ensure its continued control of the government even though many countries have refused to recognize its legitimacy. The ANC dismissed Attorney General Ortega, voted to approve its own mandate for two years, and declared itself superior to other branches of government. Ortega fled Venezuela in August 2017 and, since then, has spoken out against the Maduro government.[17] The ANC also approved a decree allowing it, not the National Assembly, to pass legislation.

From mid-2017 to May 2018, President Maduro strengthened his control over the PSUV and gained the upper hand over the MUD despite international condemnation of his actions. In October 2017, the PSUV won 18 of 23 gubernatorial elections. Although fraud likely took place given the discrepancies between opinion polls and the election results, the opposition could not prove that it was widespread.[18] There is evidence that the PSUV linked receipt of future government food assistance to votes for its candidates by placing food assistance card registration centers next to polling stations, a practice also used in subsequent elections.[19] The MUD coalition initially rejected the election results, but four victorious MUD governors took their oaths of office in front of the ANC (rather than the National Assembly), a decision that fractured the coalition.

---

[16] Venezuela's CNE reported that almost 8.1 million people voted, but a company involved in setting up the voting system alleged that the tally was inflated by at least 1 million votes. Jennifer McCoy, "Credibility of Venezuela's Electoral Process on the Line," Washington Office on Latin America, August 3, 2017.

[17] Ortega has released a video of the former head of the Brazilian construction firm Odebrecht's Venezuelan operations testifying that the company had paid President Maduro $35 million in bribes in 2013. Nicholas Casey, "Maduro is Accused in Bribery," *New York Times*, October 13, 2017; Ortega has provided evidence against other top officials to the U.S. government and evidence of government human rights abuses to the International Criminal Court. Reuters, "Venezuela Ex-Prosecutor Gives U.S. Evidence on Maduro Officials," October 13, 2017.

[18] Gideon Long, "Venezuela Elections Dogged by Fraud Claims," *Financial Times*, October 16, 2017.

[19] Michael Penfold, *Food, Technology, and Authoritarianism in Venezuela's Elections*, Woodrow Wilson Center, April 18, 2018.

With the opposition in disarray, President Maduro and the ANC moved to consolidate power and blamed U.S. sanctions for the country's economic problems. Maduro fired and arrested the head of PdVSA and the oil minister for corruption. He appointed a general with no experience in the energy sector as oil minister and head of the company, further consolidating military control over the economy. The ANC approved a law to further restrict freedom of expression and assembly.

Although most opposition parties did not participate in municipal elections held in December 2017, a few, including A New Time (UNT), led by Manuel Rosales, and Progressive Advance, led by Henri Falcón, fielded candidates. The PSUV won more than 300 of 335 mayoralties. The CNE required parties that did not participate in those elections to re-register in order to run in the 2018 presidential contest, a requirement that many of them subsequently rejected.

May 2018 Elections and Aftermath

The Venezuelan constitution established that the country's presidential elections were to be held by December 2018. Although many prominent opposition politicians had been imprisoned (Leopoldo López, under house arrest), barred from seeking office (Henrique Capriles), or in exile (Antonio Ledezma[20]) by late 2017, some MUD leaders sought to unseat Maduro through elections. Those leaders negotiated with the PSUV to try to obtain guarantees, such as a reconstituted CNE and international observers, to help ensure the elections would be as free and fair as possible. In January 2018, the ANC ignored those negotiations and called for elections to be moved up from December to May 2018, violating a constitutional requirement that elections be called with at least six months anticipation.[21] The MUD declared an election boycott, but Henri Falcón, former governor of Lara, broke with the coalition to run.

[20] Ledezma, a former mayor of metropolitan Caracas, escaped house arrest and fled to Spain in November 2017.

[21] José Ignacio Hernández G., "Rigged Elections: Venezuela's Failed Presidential Election," Electoral Integrity Project, May 30, 2018.

Venezuela's presidential election proved to be minimally competitive and took place within a climate of state repression. President Maduro and the PSUV's control over the CNE, courts, and constituent assembly weakened Falcón's ability to campaign. State media promoted government propaganda. There were no internationally accredited election monitors. The government coerced its workers to vote and placed food assistance card distribution centers next to polling stations.

The CNE reported that Maduro received 67.7% of the votes, followed by Falcón (21%) and Javier Bertucci, a little-known evangelical minister (10.8%). Voter turnout was much lower in 2018 (46%) than in 2013 (80%), perhaps due to the MUD's boycott. After independent monitors reported widespread fraud, Falcón and Bertucci called for new elections to be held.[22]

Since the May 2018 election, President Maduro has faced mounting economic problems, coup attempts, and increasing international isolation. His government has released some political prisoners, including U.S. citizen Joshua Holt, former Mayor Daniel Ceballos, opposition legislators (Gilber Caro and Renzo Prieto), and, in October 2018, former student leader Lorent Saleh. He reshuffled his Cabinet to establish Delcy Rodriguez, former head of the ANC and former foreign minister, as executive vice president in June 2018 and made additional changes within the judiciary and the intelligence services to increase his control.

With the opposition relatively weak and divided, Maduro focused on quashing coup plots and dissent within the military. His government arrested those perceived as threats, including military officers, an opposition legislator accused of involvement in an August 2018 alleged assassination attempt against Maduro, and a German journalist accused of being a spy.[23] Foro Penal and Human Rights Watch have documented several cases in which those accused of plotting coups were subjected to

---

[22] "Venezuela's Maduro Wins Boycotted Elections Amid Charges of Fraud," NPR, May 21, 2018.

[23] Andrew Rosati, "Venezuela Arrests Military Officers in Probe of Failed Assassination Bid," *Bloomberg*, August 14, 2018; Reporters Without Borders, "German Journalist Held in Venezuela Facing 28 Years in Prison on Spying Charges," December 11, 2018.

"beatings, asphyxiation and electric shocks" by the intelligence services.[24] The October 2018 death in custody of Fernando Albán, an opposition politician, provoked domestic protests and international concern.[25]

Maduro's Second Term Deemed Illegitimate; Interim Government Formed

On January 10, 2019, Maduro began a second term after winning reelection on May 20, 2018, in a contest deemed illegitimate by the opposition-controlled National Assembly and most of the international community. The United States, the European Union (EU), the Group of Seven (G-7), and most Western Hemisphere countries do not recognize the legitimacy of his mandate. They view the National Assembly as Venezuela's only democratic institution.

Under the leadership of Juan Guaidó, a 35-year-old industrial engineer from the VP party who was elected president of the National Assembly on January 5, 2019, the opposition has been reenergized. In mid-January, Guaidó announced he was willing to serve as interim president until new presidential elections are held. Buoyed by the massive turnout for protests he called for, Guaidó took the oath of office on January 23, 2019. Since then, the National Assembly has enacted resolutions declaring Maduro's mandate illegitimate, establishing a framework for a transition government, drafting a proposal to offer amnesty for officials who support the transition, and creating a strategy for receiving humanitarian assistance.

Guaidó and his supporters have organized two high-profile efforts to encourage security (military and police) forces to abandon Maduro, neither of which has succeeded.[26] On February 23, they sought to bring emergency supplies into the country that had been positioned on the Colombia- and

---

[24] Human Rights Watch, "Venezuela: Suspected Plotters Tortured," January 9, 2019.

[25] The government maintains that Albán committed suicide; U.N. officials are investigating his death as a murder. Associated Press, "Fernando Albán: UN to Investigate Death in Custody of Venezuelan Politician," October 9, 2018.

[26] Mery Mogollan and Patrick J. McDonnell, "In Venezuela, U.S.-Guaido Strategy Flops Again: Is This Working?" *Los Angeles Times*, May 5, 2019.

Brazil-Venezuela borders by the United States and other donors. Venezuelan national guard troops and armed civilian militias (*colectivos*) loyal to Maduro killed seven individuals and injured hundreds as they prevented the aid convoys from crossing the border.[27] The rest of the military did not attack civilians. While that aid remains blocked, U.N. agencies are ramping up their efforts in Venezuela, and both Guaidó and Maduro agreed in late March to allow the International Federation of Red Cross and Red Crescent Societies (IFRC) to begin providing assistance in Venezuela (see "International Response Inside Venezuela" below).[28]

By April 2019, more than 1,000 Venezuelan security forces (mostly from the national guard and police) had defected into Colombia. The Colombian military has disarmed and placed them in hotels near the border along with their family members.[29] In May 2019, Colombia's migration agency signed an agreement with the interim government of Venezuela to grant security forces that have defected from Maduro temporary legal status to work and receive assistance from the Colombian government.[30] Prospects for those who have defected is reportedly difficult, as they have become among the more than 1.2 million Venezuelans in Colombia.[31]

On April 30, 2019, Guaidó and Leopoldo López, head of the VP party who had been released from house arrest by pro-Guaidó military officials, called for a civil-military rebellion against Maduro. National guard and militias loyal to Maduro violently put down pro-Guaidó supporters and attacked journalists. Social media accounts and independent media were blocked.[32]

---

[27] Foro Penal, *Report on Repression in Venezuela*, February 2019.

[28] IFRC, "IFRC to Bring Humanitarian Aid into Venezuela," March 29, 2019. See https://media.ifrc.org/ifrc/pressrelease/ifrc-bring-humanitarian-aid-venezuela.

[29] Karen DeYoung and Mary Beth Sheridan, "Venezuelan Military Foils U.S. Hopes," *Washington Post*, April 14, 2019. The article states that there are more than 2,000 troops and family members from Venezuelan waiting in border area hotels.

[30] Gobierno de Colombia, "Colombia Determina Esquema de Atención para Ex-Militares y Ex-Policias Venezolanos que se Encuentran en el Territorio Nacional," May 15, 2019.

[31] Solsvik Terje and Angus Berwick, "Venezuela's Guaidó Vows Protests as Oslo Talks Produce No Deal," Reuters, May 29, 2019.

[32] International Crisis Group, *Picking up the Pieces After Venezuela's Quashed Uprising*, May 1, 2019.

U.S. officials asserted that top Maduro officials, including the head of the Supreme Court and the defense minister, had pledged to support the uprising but changed their minds.[33] As the day ended, López and other opposition lawmakers sought refuge in various foreign embassies.

Guaidó and his allies continue to hope that sustained protests and international pressure will lead to military defections that will compel Maduro to leave office and allow elections to be held. Nevertheless, they have also sent envoys to Norway for talks on a political solution to the crisis.[34] Maduro's harassment of the National Assembly and the Maduro-aligned Supreme Court's efforts to arrest and prosecute lawmakers involved in the uprising could derail those efforts.

Many observers regard the military's participation as essential for the opposition's transition plan to work. Aside from the former head of the national intelligence agency (General Manuel Christopher Figuera), who supported the uprising on April 30, the military high command remains loyal to Maduro. Top military leaders have enriched themselves through corruption, drug trafficking, and other illicit industries (see "Organized Crime-Related Issues" below).[35] This illicit income, they argue, has blunted the impact of U.S. sanctions on Maduro's closest allies.[36] Some military leaders may fear that they could face prosecution for human rights abuses under a new government, even though the opposition has proposed amnesty for those who join their side. The U.S. government has also said that it may remove sanctions on officials who abandon Maduro and side with Guaidó (as they did with General Figuera), but that process could be more difficult, depending upon the individual and type of sanctions involved.[37]

---

[33] "Bolton Speaks About Venezuela," *Washington Post*, April 30, 2019.

[34] Terje and Berwick, "Venezuela's Guaidó Vows Protests."

[35] Douglas Farah and Caitlyn Yates, *Maduro's Last Stand: Venezuela's Survival through the Bolivarian Joint Criminal Enterprise*, IBI Consultants and the National Defense University, May 2019.

[36] Farah and Yates, *Maduro's Last Stand.*

[37] The White House, "Remarks by Vice President Pence at the Washington Conference on the Americas," May 7, 2019; Reuters, "Relief from U.S. Sanctions Will Not Come Easily for Some Venezuelans," May 30, 2019.

HUMAN RIGHTS

Human rights organizations and U.S. officials have expressed concerns for more than a decade about the deterioration of democratic institutions and threats to freedom of speech and press in Venezuela.[38] Human rights conditions in Venezuela have deteriorated even more under President Maduro than under former President Chávez. Abuses have increased, as security forces and allied armed civilian militias (*colectivos*) have been deployed to violently quash protests. In August 2017, the United Nations Office of the High Commissioner for Human Rights (UNOCHR) issued a report on human rights violations perpetrated by the Venezuelan security forces against protestors earlier that year. According to the report, credible and consistent accounts indicated that "security forces systematically used excessive force to deter demonstrations, crush dissent, and instill fear."[39] The U.N. report maintained that many of those detained were subject to cruel, degrading treatment and that in several cases, the ill treatment amounted to torture.

In June 2018, UNOCHR issued another report documenting abuses committed by units involved in crime fighting, the scale of the health and food crisis, and the continued impunity in cases involving security officers who allegedly killed people during the protests.[40] Selected human rights reports from 2018 to 2019 include the following:

- In June 2018, PROVEA, one of Venezuela's leading human rights organizations, asserted that 2017 was the worst year for human

---

[38] For background, see CRS Report R43239, *Venezuela: Issues for Congress, 2013-2016*, by Mark P. Sullivan.

[39] Office of the United Nations High Commissioner for Human Rights (UNOCHR), *Human Rights Violations and Abuses in the Context of Protests in the Bolivarian Republic of Venezuela from 1 April to 31 July 2017*, August 30, 2017. The Venezuelan human rights group Foro Penal and Human Rights Watch maintain that more than 5,300 Venezuelans were detained during the protests. They documented inhumane treatment of more than 300 detainees. Human Rights Watch, Foro Penal, *Crackdown on Dissent: Brutality, Torture, and Political Persecution in Venezuela*, November 2017.

[40] UNOCHR, *Venezuela: Continued Impunity amid Dismal Human Rights Situation*, June 22, 2018.

rights in Venezuela since the report was first published in 1989. In addition to violating political and civil rights, PROVEA denounces the Maduro government's failure to address the country's humanitarian crisis, citing its "official indolence" as causing increasing deaths and massive emigration.[41]

- In May 2018, an independent panel of human rights experts added a legal assessment to a report containing information and witness testimonies gathered by the Organization of American States (OAS) recommending that the International Criminal Court (ICC) investigate credible reports that the Venezuelan government committed crimes against humanity.[42]
- In February 2019, the Venezuelan Human Rights group Foro Penal documented seven deaths, 107 arbitrary detentions, and 58 bullet injuries that resulted from the use of force by state security forces and *colectivos* that blocked aid from entering the country on February 22-23, 2019.[43]
- In March 2019, the U.S. State Department's *Country Report on Human Rights Practices for 2018* cited "extrajudicial killings by security forces, including *colectivos*; torture by security forces; harsh and life-threatening prison conditions; and political prisoners" as the most serious human rights abuses in Venezuela. According to the report, the government did not investigate officials alleged to have committed abuses amidst widespread impunity.[44]

Analysts predict increasing repression, as Maduro has called for the arrest of Leopoldo López and opposition lawmakers involved in the April 30 uprising. The Supreme Court has stripped several lawmakers' immunity

[41] Programa Venezolana de Educación y Acción en Derechos Humanos (PROVEA), *Informe Anual Enero-Diciembre 2017*, June 2018.

[42] Organization of American States (OAS), Report of the General Secretariat of the OAS and the Panel of Independent International Experts on the Possible Commission of Crimes Against Humanity in Venezuela, May 29, 2018.

[43] Foro Penal, *Report on Repression in Venezuela*. February 2019.

[44] U.S. Department of State, *2018 Country Reports on Human Rights Practices: Venezuela*, March 13, 2019.

from prosecution, increasing the likelihood that they may be arrested and possibly tortured if they do not seek protection in a foreign embassy or go into exile. Maduro's intelligence police is currently holding and reportedly torturing Juan Requesens (a legislator detained in August 2018), Roberto Marrero (Guaidó's chief of staff detained in March 2019), and Edgar Zambrano (the vice president of the National Assembly detained May 9).[45] Some fear that Guaidó himself could face arrest or exile.

For other sources on human rights in Venezuela, see Appendix B.

Investigations into Human Rights Abuses

In September 2017, several countries urged the U.N. Human Rights Council to support the High Commissioner's call for an international investigation into the abuses described in the U.N.'s August 2017 report on Venezuela.[46] In June 2018, the High Commissioner for Human Rights urged the U.N. Human Rights Council to launch a commission of inquiry to investigate the abuses it documented in that and a follow-up report. It referred the report to the prosecutor of the ICC. On September 26, 2018, the U.N. Human Rights Council adopted a resolution on Venezuela expressing "its deepest concern" about the serious human rights violations described in the June 2018 report, calling upon the Venezuelan government to accept humanitarian assistance and requiring a UNOCHR investigation on the situation in Venezuela to be presented in 2019. In March 2019, a technical team from UNOCHR visited Venezuela, possibly paving the way for a visit by High Commissioner Michelle Bachelet.

In addition to the UNOCHR, former Venezuelan officials, the OAS, and neighboring countries have asked the ICC to investigate serious human rights violations committed by the Maduro government; the ICC

[45] OAS, "IACHR Condemns the Escalation of Attacks Against Members of the Venezuelan National Assembly," August 29, 2018.

[46] Tamara Taraciuk Broner, "U.N. Human Rights Council Confronts Venezuela," Human Rights Watch, October 2, 2017; UNOCHR, "Venezuela: Continued Impunity Amid Dismal Human Rights Situation - UN report," June 22, 2018; UNOCHR, "Human Rights Council Adopts 10 Resolutions and one Presidential Statement," September 27, 2018.

prosecutor opened a preliminary investigation in February 2018.[47] In November 2017, former Attorney General Luisa Ortega presented a dossier of evidence to the ICC that the police and military may have committed more than 1,800 extrajudicial killings as of June 2017. In the dossier, Ortega urged the ICC to charge Maduro and several officials in his Cabinet with serious human rights abuses. An exiled judge appointed by the National Assembly to serve on the "parallel" supreme court of justice also accused senior Maduro officials of systemic human rights abuses before the ICC. On September 26, 2018, the governments of Argentina, Canada, Chile, Colombia, Paraguay, and Peru requested an investigation of Venezuela's actions by the ICC—the first time fellow states party to the Rome Statute asked for an investigation into the situation of another treaty member.

ECONOMIC CRISIS[48]

For decades, Venezuela was one of South America's most prosperous countries. Venezuela has the world's largest proven reserves of oil, and its economy is built on oil. (See also "Energy Sector Concerns and U.S. Economic Sanctions" below.) Oil traditionally has accounted for more than 90% of Venezuelan exports, and oil sales have funded the government budget. Venezuela benefited from the boom in oil prices during the 2000s. President Chávez used the oil windfall to spend heavily on social programs and expand subsidies for food and energy, and government debt more than doubled as a share of gross domestic product (GDP) between 2000 and

---

47 "Venezuela's Ex-prosecutor Urges ICC to Probe Maduro," AFP, November 16, 2017; Antonio Maria Delgado, "Venezuelan Officials Accused of Crimes Against Humanity in The Hague," *Miami Herald*, November 21, 2017; ICC, "Statement of the Prosecutor of the International Criminal Court, Mrs Fatou Bensouda, on opening Preliminary Examinations into the situations in the Philippines and in Venezuela; February 8, 2018; ICC, "Statement of the Prosecutor of the International Criminal Court, Mrs Fatou Bensouda, on the Referral by a Group of Six States Parties Regarding the Situation in Venezuela," September 27, 2018.

48 This section was authored by Rebecca M.Nelson, Specialist in International Trade and Finance. For more information, see CRS Report R45072, *Venezuela's Economic Crisis: Issues for Congress*, by Rebecca M. Nelson.

2012.[49] Chávez also used oil to expand influence abroad through *PetroCaribe*, a program that allowed Caribbean Basin countries to purchase oil at below-market prices.

Although substantial government outlays on social programs helped Chávez curry political favor and reduce poverty, economic mismanagement had long-term consequences. Chávez moved the economy in a less market-oriented direction, with widespread expropriations and nationalizations, as well as currency and price controls. These policies discouraged foreign investment and created market distortions. Government spending was not directed toward investment to increase economic productivity or diversify the economy from its reliance on oil. Corruption proliferated.

When Nicolás Maduro took office in 2013, he inherited economic policies reliant on proceeds from oil exports. When world oil prices crashed by nearly 50% in 2014, the Maduro government was ill-equipped to soften the blow. The fall in oil prices strained public finances. Instead of adjusting fiscal policies through tax increases and spending cuts, the Maduro government tried to address its growing budget deficit by printing money, which led to inflation. The government also tried to curb inflation through price controls, although these controls were largely ineffective in restricting prices, as supplies dried up and transactions moved to the black market.[50]

Meanwhile, the government continued to face a substantial debt burden, with debt owed to private bondholders, China, Russia, multilateral lenders, importers, and service companies in the oil industry. Initially, the government tried to service its debt, fearing legal challenges from bondholders. To service its debt, it cut imports, including of food and medicine, among other measures. By late 2017, the government had largely stopped paying its bondholders, and Maduro announced plans to restructure its debt with private creditors. It also restructured its debt with Russia in 2017.[51] The Trump Administration has increased economic pressure on the Maduro government through sanctions. (See "U.S.

[49] *The Economist*, "How Chávez and Maduro Have Impoverished Venezuela," April 6, 2017.
[50] *The Economist*, "How Chávez and Maduro Have Impoverished Venezuela."
[51] John Paul Rathbone and Robin Wigglesworth, "Venezuela Debt: US, Russia, and China Play for High Stakes," *Financial Times*, November 22, 2017.

Sanctions on Venezuela.") For example, in August 2017, it restricted Venezuela's ability to access U.S. financial markets, which exacerbated the government's fiscal situation, and in January 2019, it imposed new restrictions on Venezuela's oil sector.

Over the past several years, Venezuela's economy has collapsed, and it is difficult to overstate the extent of its challenges. According to some economists, it is the single largest economic collapse outside of war in at least 45 years.[52] According to forecasts by the International Monetary Fund (IMF), Venezuela's economy in 2019 will contract for the sixth year in a row (Figure 2).[53] The IMF projects that Venezuela's gross domestic product in U.S. dollars fell from $331 billion in 2012 to $98 billion in 2018 and will fall again to $76 billion in 2019 (Figure 2).

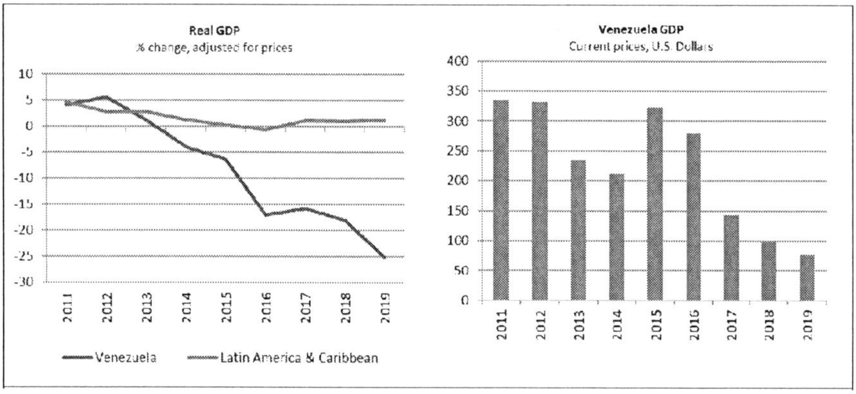

Source: IMF, *World Economic Outlook*, April 2019.
Notes: 2019 data are forecasts.

Figure 2. Venezuela's Economy: Collapsing GDP.

Hyperinflation has devastated the economy. The Maduro government rapidly expanded the money supply to finance budget deficits, which led to one of the worst cases of hyperinflation in history, comparable to Germany

52 Anatoly Kurmanaev, "Venezuela's Collapse Is the Worst Outside of War in Decades, Economists Say," *New York Times*, May 17, 2019.

53 IMF, *World Economic Outlook*, April 2019.

in 1923 or Zimbabwe in the late 2000s.[54] The IMF forecasts that inflation in Venezuela will reach 10 million percent in 2019.[55] Hyperinflation, as well as low foreign exchange reserves, which make it difficult for Venezuela to import goods and services, has created shortages of critical supplies (including food and medicine), leading to a humanitarian disaster and fueling massive migration (see "Humanitarian Situation" below).

Despite pledges to restructure the country's debt, the government has made no progress in negotiations with private creditors and the country remains in default on most of its sizeable debt. The government has continued payments on a few select debts, including loans from Russia, a geopolitically and financially important supporter for the Maduro government, and some bonds held by private creditors that, if defaulted on, could result in legal challenges by investors and seizure of U.S. refiner Citgo (owned by PdVSA). Meanwhile, the government continues to run large budget deficits, forecast at 30% of GDP in 2018 and 2019, amid high debt levels (estimated to reach 214% of GDP in 2019).[56] By one measure, debt relative to exports, Venezuela is the world's most heavily indebted country.[57]

In general, the government has been slow to address the economic crisis or acknowledge the government's role in creating it. Instead, the government has largely blamed the country's struggles on a foreign "economic war," a thinly veiled reference to U.S. sanctions.[58] In February 2018, as a way to raise new funds, the cash-strapped government launched a new digital currency, the "petro," backed by oil and other commodities, although there are serious questions about the petro's operational viability.[59] In August 2018, the government acknowledged, for the first

---

[54] For example, see Alejandro Werner, "Outlook for the Americas: A Tougher Recovery," IMF Blog, July 23, 2018.

[55] IMF, *World Economic Outlook*, October 2018. Percent change in GDP in constant dollars (adjusted for inflation).

[56] IMF, *World Economic Outlook*, October 2018.

[57] Interview of Ricardo Hausmann by Sebastian Pellejero, "Can Venezuela Resurrect Its Economy?," Council on Foreign Relations, July 26, 2017.

[58] Rachelle Krygier, "After Years of Crisis, Venezuela's Maduro Might Finally Be Ready to Accept Some Help," *Washington Post*, December 12, 2018.

[59] See CRS In Focus IF10825, *Digital Currencies: Sanctions Evasion Risks*, by Rebecca M. Nelson and Liana W. Rosen; Eric Lam, "Here's What Maduro Has Said of Venezuela's

time, its role in creating hyperinflation and announced a new set of policies for addressing the economic crisis, such as introducing a new "sovereign bolívar," which removed five zeros from the previous currency (the bolívar); cutting the government budget deficit from 30% in 2018 to zero; speeding up tax collection; and increasing the minimum salary by more than 3,000%.[60] However, the government has been slow to enact even these modest reforms, and Maduro has offered few new proposals since starting his second term.[61]

Venezuela's economic situation has become more difficult following the rollout of new U.S. oil sanctions in January 2019, although it has worked with Russia, Turkey, and other countries to mitigate the sanctions' impact.[62] The Venezuelan government has turned to selling gold reserves in a desperate attempt to raise funds, with nearly $800 million sold to Turkey in 2018.[63] As the United States has since sanctioned Venezuela's central bank and state-run gold mining company, analysts maintain that Maduro will likely increase illicit gold shipments sent through neighboring countries, a key revenue stream for the military.[64]

Venezuela's economic crisis has been ongoing for a number of years, and the outlook is bleak. There is neither a clear nor a quick resolution on the horizon, particularly given the concurrent political crisis. The government's policy responses to the economic crisis have been widely criticized as inadequate. The government appears loathe to adopt policies widely viewed by economists as necessary to restoring the economy: removing price controls, creating an independent central bank, entering an IMF program that could unlock broader international assistance, and restructuring its debt with private bondholders.

Petro Cryptocurrency," *Bloomberg*, August 20, 2018; Aaron Mak, "What Does It Mean for Venezuela to Peg Its New Currency to a Cryptocurrency?," *Slate*, August 22, 2018; and Brian Ellsworth, "Special Report: In Venezuela, New Cryptocurrency Is Nowhere to Be Found," *Retuers*, August 30, 2018.

[60] "Nicolas Maduro Tries to Rescue Venezuela's Economy," *Economist*, August 23, 2018.

[61] Vivian Sequera, "Venezuela's Maduro Offers Few Fresh Ideas as Economy Circles Drain," Reuters, January 14, 2019.

[62] Anatoly Kurmanaev and Clifford Krauss, "U.S. Sanctions Are Aimed at Venezuela's Oil. Its Citizens May Suffer First," *New York Times*, February 8, 2019.

[63] Reuters, "Venezuela Exported $779 Million in Gold to Turkey in 2018," July 19, 2018.

[64] International Crisis Group, *Gold and Grief in Venezuela's Violence South*, February 28, 2019.

The Maduro government is particularly resistant to assistance from the IMF, which would be a key component of any multi-donor international financial assistance package. Venezuela has not allowed the IMF to conduct routine surveillance of its economy since 2004, and the IMF has found the government in violation of its commitments as an IMF member. Although the Venezuelan government provided it with some economic data as required by all IMF members in late 2018, it has hardly been a turning point in the Maduro government's willingness to engage with the international organization. Some analysts believe major change in Venezuela's overall economic strategy will come only if and when there is a change in government. Even then, analysts have projected it could take several years to address, including what is expected to be the most complex sovereign debt restructuring in history and an international financial assistance package of $100 billion to help the country's economy to recover.[65]

HUMANITARIAN SITUATION[66]

Before the political upheaval in Venezuela began in January 2019, Venezuelans were already facing a humanitarian crisis due to a lack of food, medicine, and access to social services. Political persecution, hyperinflation, loss of income, and oppressive poverty also contributed to a dire situation. According to household surveys, the percentage of Venezuelans living in poverty increased from 48.4% in 2014 to 87% in 2017.[67] U.N. officials estimate that approximately 7 million people are in need of humanitarian assistance, with pregnant and nursing women, those with chronic illnesses, indigenous people, migrants, children under five,

---

[65] For example, see Colby Smith, "Bondholders Brace for Venezuelan Regime Change," *Financial Times*, February1, 2019; Sam Fleming and James Politi, "US Considers Emergency Economic Aid for Venezuela," *Financial Times*, March 3, 2019.

[66] For more information, see CRS In Focus IF11029, *The Venezuela Regional Migration Crisis*, by Rhoda Margesson and Clare Ribando Seelke.

[67] Universidad Católica Andrés Bello, *Encuesta Sobre Condiciones de Vida:Venezuela 2017*, February 2018.

and people with disabilities particularly in need.[68] Successive electrical blackouts in March 2019 made conditions worse inside Venezuela, limiting people's access to power and clean water and further contributing to health risks. While power has been restored in many parts of the country, rationing in many states continues to take place.[69]

As of April 2019, the U.N. High Commissioner for Refugees (UNHCR) and the International Organization for Migration (IOM) estimated that approximately 3.7 million Venezuelans (roughly

12% of the population) had left the country (see Figure 3). Of those, most (an estimated 3 million Venezuelans) remained in Latin America and the Caribbean, including in Colombia (1.2 million), Peru (728,000), Chile (288,000), Ecuador (221,000), Argentina (130,000), Brazil (96,000), and Panama (94,000).[70] UNHCR and IOM estimate that by the end of 2019, the number of

Venezuelan refugees and migrants could reach over 5.3 million.[71] Although UNHCR does not consider most Venezuelans to be refugees, it asserts that a significant number need humanitarian assistance, international protection, and opportunities to regularize their status.

The crisis in Venezuela is affecting the entire region. Neighboring countries, particularly Colombia, are straining to absorb arrivals often malnourished and in poor health.[72] The spread of previously eradicated diseases, such as measles, is also a major concern.[73] Responses to the

---

[68] U.N. Under-Secretary-General for Humanitarian Affairs and Emergency Relief Coordinator Mark Lowcock, "Briefing to the Security Council on the Humanitarian Situation in Venezuela," April 10, 2019.

[69] U.N. Office for the Coordination of Humanitarian Affairs, "Venezuela—UN Humanitarian Scale-Up Situation Report, January-April 2019," May 2019.

[70] U.N. Coordination Platform for Refugees and Migrants from Venezuela, "Latin America and the Caribbean: Venezuelan Refugees and Migrants in the Region," April 2019; U.N. Coordination Platform for Refugees and Migrants from Venezuela, "Latin America and the Caribbean: Venezuelan Refugees and Migrants in the Region," March 2019. See also UNHCR, "Americas Monthly Report," March 2019. According to the Brazilian government, the country has "received" some 200,000 Venezuelans, but not all of those individuals have remained in Brazil. UNHCR, "Americas Monthly Report," March 2019.

[71] USAID, *Venezuela Regional Crisis*, FY2019, April 10, 2019.

[72] The spread of previously eradicated diseases, such as measles, is a major regional concern.

[73] Human Rights Watch and John Hopkins School of Public Health, *Venezuela's Humanitarian Emergency: Large-Scale UN Response Needed to Address Health and Food Crises*, April 2019.

Venezuelan arrivals vary by country and continue to evolve with events on the ground. Humanitarian experts are most concerned about the roughly 60% of Venezuelans in neighboring countries who lack identification documents, which makes them vulnerable to arrest and deportation by governments and to abuse by criminal groups, including human trafficking.[74]

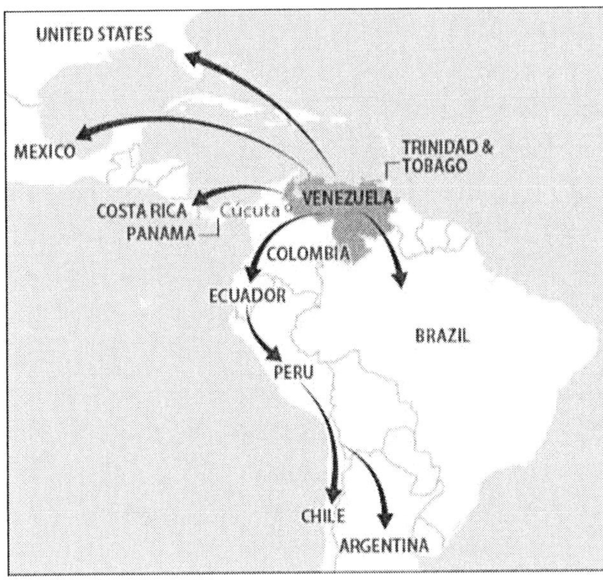

Source: UNHCR.

Figure 3. Venezuelan Migrants and Asylum Seekers: Flows to the Region and Beyond.

Venezuela's exodus is an unprecedented displacement crisis for the Western Hemisphere, which has in place some of the highest protection standards in the world for displaced and vulnerable persons. Countries in the region are under pressure to examine their respective migration and asylum policies and to address, as a region, the legal status of Venezuelans who have fled their country. It remains to be seen how events unfolding inside Venezuela will continue to the impact the ongoing humanitarian situation and response.

---

[74] The Venezuelan government has made it increasingly difficult for Venezuelans to obtain a valid passport and therefore legal status outside the country.

International Response Inside Venezuela

In 2018, Maduro publicly rejected offers of international aid, although some humanitarian assistance was provided.[75] Since then, U.N. humanitarian entities and partners have made progress in scaling up their humanitarian and protection activities. The number of U.N. staff in the country has doubled since 2017, and the U.N. Office for the Coordination of Humanitarian Affairs (UNOCHA) has set up regional hubs in the country. The Cooperation and Assistance Coordination Team, led by the U.N. Resident Coordinator, was established in February 2019 to facilitate humanitarian coordination in Venezuela. A needs overview for Venezuela was completed in March and indicated significant humanitarian needs exist across sectors. Based on this data, UNOCHA and other humanitarian partners are developing a Humanitarian Response Plan for Venezuela, which will focus on six sectors and two sub-sectors: health; food security and agriculture; nutrition; water, sanitation, and hygiene; protection (including child protection and gender-based violence); and education.[76]

In addition, on March 29, 2019, IFRC announced plans to scale up humanitarian activities inside Venezuela in coordination with the Venezuelan Red Cross and with the agreement of both the interim government and Maduro.[77] On April 11, the International Committee of the Red Cross (ICRC), which has been present in Venezuela since 1966, announced an expansion of its work on four pressing humanitarian issues: migration, health, water and sanitation, and detention.[78]

---

[75] For example, in November 2018, the U.N. Central Emergency Response Fund allocated $9.2 million for Venezuela through U.N. entities. The World Health Organization has helped provide millions of vaccines against measles, mumps, and rubella.

[76] UNOCHA, "Venezuela—UN Humanitarian Scale-Up Situation Report, January-April 2019," May 2019.

[77] IFRC, "IFRC to Bring Humanitarian Aid into Venezuela," March 29, 2019, https://media.ifrc.org/ifrc/press-release/ ifrc-bring-humanitarian-aid-venezuela.

[78] ICRC, "Venezuela: ICRC to Expand Humanitarian Effort," April 11, 2019. As part of the Red Cross Movement, and in keeping with their mandates, both IFRC and ICRC provide humanitarian assistance in an independent, neutral, impartial, and unhindered manner.

International Humanitarian Response in the Region

The U.N. Secretary-General appointed UNHCR and IOM to coordinate the international response to the needs of displaced Venezuelans and host communities in the region, which includes governments, U.N. entities, nongovernmental organizations (national and international), the Red Cross Movement, faith-based organizations, and civil society in the region. The U.S. government is also providing humanitarian assistance and helping to coordinate regional response efforts. (See "U.S. Humanitarian and Related Assistance," below.) Former Guatemalan Vice President Eduardo Stein, the U.N. Joint Special Representative for Venezuelan Refugees and Migrants, is charged with promoting dialogue and consensus in the region and beyond on the humanitarian response.

In mid-December 2018, UNHCR and IOM launched the Regional Refugee and Migrant Response Plan, which is the first of its kind in the Americas: an operational and coordination strategy and an appeal for $738 million in funding to support over 2 million Venezuelans in the region and half a million people in host communities.[79] According to the Financial Tracking Service, as of May 24, 2019, the appeal was 20.7% funded.[80] Protection and assistance needs are significant for arrivals and host communities. Services provided vary by country but include support for reception centers and options for shelter; emergency relief items, such as emergency food assistance, safe drinking water, and hygiene supplies; legal assistance with asylum applications and other matters; protection from violence and exploitation; and the creation of temporary work programs and education opportunities.

---

[79] The plan includes U.N. entities, nongovernmental organizations, faith-based organizations, and civil society and is in cooperation with the Red Cross Movement. See *Coordination Platform for Response for Venezuelans, Regional Refugee and Migrant Response Plan for Refugees and Migrants from Venezuela*, January-December 2019, published December 14, 2018. It was put together by 95 organizations covering 16 countries. Organizations have also launched separate appeals.

[80] See Financial Tracking Service at https://fts.unocha.org/appeals/726/summary.

FOREIGN INVOLVEMENT IN VENEZUELA'S POLITICAL CRISIS[81]

The international community remains divided over how to respond to the political crisis in Venezuela. One group of countries supports the Guaidó government, another supports Maduro, and a third group of countries—including Mexico, Norway, Uruguay, and some Caribbean nations—has remained neutral in the crisis.

These divisions have thus far stymied action by the U.N. and the OAS to help facilitate a political solution to Venezuela's crisis. Within the U.N. Security Council, Russia and (to a lesser extent) China support Maduro. The United States, France, and the United Kingdom (UK) support Guiadó. At the OAS, recent resolutions have amassed enough votes (19 of 34 member states) to declare Maduro's 2018 reelection illegitimate.[82] The OAS remains divided, however, over what role it should play in hastening a return to democracy in Venezuela. Countries in the Caribbean Community (CARICOM), many of which used to receive subsidized oil from Venezuela, have been particularly reluctant to intervene in Venezuela's internal affairs.

The United States and 53 other countries have formally recognized the interim government of Juan Guaidó. These countries include most members of the EU, Canada, 14 countries in Latin America and the Caribbean,

---

[81] See also CRS In Focus IF11216, *Venezuela: International Efforts to Resolve the Political Crisis*, by Clare Ribando Seelke.

[82] The OAS requires 18 votes to pass a resolution of the Permanent Council. In June 2018, 19 of 34 member states passed a resolution stating that the May 2018 presidential election in Venezuela lacked legitimacy and authorizing countries to take measures, including sanctions, necessary to hasten a return to democracy. In January 2019, the same 19 states approved a resolution that refused to recognize the legitimacy of Maduro's second term; called for new presidential elections; and urged all member states to adopt diplomatic, political, economic, and financial measures to facilitate the prompt restoration of the democratic order in Venezuela. Those countries include Argentina, the Bahamas, Barbados, Brazil, Canada, Chile, Colombia, Costa Rica, the Dominican Republic, Honduras, Jamaica, Guatemala, Guyana, Mexico, Panama, Paraguay, Peru, St. Lucia, and the United States. The four countries that voted against the January 2019 resolution were Bolivia, Dominica, St. Vincent and the Grenadines, and Venezuela. The 11 countries that abstained included Antigua and Barbuda, Belize, Ecuador, Grenada, El Salvador, Haiti, Nicaragua, St. Kitts and Nevis, Suriname, Trinidad and Tobago, and Uruguay.

Australia, Israel, Japan, and South Korea, among others. Some of those countries have undertaken initiatives to increase pressure on Maduro to leave office. Canada, the EU, Panama, and Switzerland have placed targeted sanctions on Maduro officials and have frozen reserves formerly controlled by the Maduro government and/or blocked suspicious financial transactions involving Maduro.[83] Other countries have withdrawn their diplomats from Caracas and/or accepted the credentials of diplomats representing the Guaidó government.

While they back a transition leading to Maduro leaving power, the EU and many Western Hemisphere countries (including Argentina, Chile, Colombia, Brazil, and Peru) have opposed the prospect of U.S. military intervention in Venezuela.[84] They and others fear the regional implications that instability and violence spawned by the use of military force could cause, particularly given the armed civilian militias and illegally armed groups in Venezuela, many of which are not under Maduro's control.[85] Among countries that back a political transition in Venezuela that involves Maduro leaving office, two groups have emerged: the Lima Group and the International Contact Group.

Lima Group

On August 8, 2017, 12 Western Hemisphere countries signed the Lima Accord, a document rejecting what it described as the rupture of democracy and systemic human rights violations in Venezuela.[86] The

---

[83] Callum Burroughs, "The Bank of England Has Refused to Give Back $1.56 Billion in Venezuelan Gold after Countries Around the World Say the Regime Is Illegitimate," *Business Insider*, February 6, 2019.

[84] European Council, "Declaration by the High Representative on Behalf of the EU on the Latest Events in Venezuela," Government of Canada, Lima Group Declaration, February 25, 2019.

[85] Testimony of Cynthia J. Arnson before the Senate Foreign Relations Subcommittee on the Western Hemisphere, Transnational Crime, Civilian Security, Democracy, Human Rights, and Global Women's Issues, hearing on "U.S.- Venezuela Relations and Path to a Democratic Transition," March 7, 2019.

[86] Government of Canada, "Lima Group Declaration," August 8, 2017. These statements are available in English at https://www.international.gc.ca/world-monde/issues_development-

signatory countries included Argentina, Brazil, Canada, Chile, Colombia, Costa Rica, Guatemala, Honduras, Mexico, Panama, Paraguay, and Peru. In 2018, Guyana and St. Lucia joined the Lima Group, which did not recognize Maduro's May 2018 reelection.

On January 4, 2019, 13 members of the Lima Group signed a declaration urging President Maduro not to assume power on January 10, 2019.[87] The countries resolved to reassess their level of diplomatic engagement with Venezuela, implement travel bans or sanctions on Maduro officials (as Canada and Panama have), suspend military cooperation with Venezuela, and urge others in the international community to take similar actions. Under leftist President Andrés Manuel López Obrador, Mexico is no longer participating in the Lima Group. El Salvador is likely to join the Lima Group after President-elect Nayib Bukele's June 1 inauguration.

The group has denounced human rights violations by the Maduro government and urged Venezuelan armed forces to demonstrate their loyalty to Guaidó but opposed U.S. or regional military intervention in the crisis. On May 3, 2019, the Lima Group issued a declaration signed by 12 countries (excluding St. Lucia and Guyana but including the interim government of Venezuela) asking the ICG to meet to coordinate efforts and pledging to seek Cuba's help in resolving Venezuela's crisis.[88] Representatives from the Lima Group and the ICG met at the ministerial level on June 3.

International Contact Group (ICG)

The EU-backed ICG—composed of Costa Rica, Ecuador, France, Germany, Italy, the Netherlands, Portugal, Spain, Sweden, the United Kingdom, and Uruguay—first convened on February 7, 2019. The group

---

enjeux_developpement/response_conflictreponse_conflits/crisis-crises/venezuela.aspx?lang=eng.

[87] Government of Canada, "Lima Group Declaration," January 4, 2019.

[88] Government of Canada, "Lima Group Declaration," May 3, 2019.

aimed to "establish necessary guarantees for a credible electoral process, within the earliest time frame possible" and to hasten the delivery of humanitarian aid into Venezuela.[89] U.S. officials have expressed skepticism at the proposal, noting that Maduro has used past attempts at dialogue (brokered by the Vatican and others) as a delaying tactic. ICG supporters maintain that the "necessary guarantees" include naming a new electoral council, releasing political prisoners, and ending all bans on political parties and candidates.[90]

Since February, the ICG has met twice, most recently on May 6-7 in Costa Rica. At that meeting, also attended by Chile from the Lima Group and representatives from CARICOM and the Vatican, the ICG decided to send a political mission to Caracas (which they did) and to work with the Lima Group to hasten a political solution to the crisis.[91] The ICG backs the dialogue process (discussed below) that has gotten underway in Norway.[92] China has pledged to work with the EU and others for a political settlement to the Venezuela crisis.[93]

Cuba, Russia, and China

Within Latin America, Cuba, Nicaragua, Bolivia, and a few other countries have provided various kinds of support to Maduro and have sought to block anti-Maduro actions within international organizations. Among Maduro's supporters in Latin America, the most significant is Cuba. Cuba's close relationship with Venezuela was solidified in 2000, when the countries signed an agreement for Venezuela to provide Cuba oil in exchange for technical assistance and other services. Estimates of the

---

[89] EU, "International Contact Group—Meeting 7 February," February 7, 2019.

[90] David Smilde and Geoff Ramsey, *The Fraught Path Forward: Venezuela and the International Contact Group*, Washington Office on Latin America, March 5, 2019.

[91] EU, "Venezuela: Statement from the Ministerial Meeting of the International Contact Group," May 7, 2019.

[92] EU, "Statement by the International Contact Group on the Norwegian Facilitated Negotiation Process," May 26, 2019.

[93] Ministry of Foreign Affairs of the People's Republic of China," Foreign Ministry Spokesperson Geng Shuang's Regular Press Conference on May 8, 2019," May 8, 2019.

number of Cuban personnel in Venezuela vary, but a 2014 study estimated that there were 40,000, 75% of whom were health care workers.[94] It is unclear how many of those professionals have stayed in the country, but Cuban intelligence officers have reportedly helped the Maduro government identify and disrupt coup plots, a role that the Trump Administration has frequently denounced.[95] Although Cuba has imported oil from other sources to make up for dwindling Venezuelan supplies, the Maduro government remains committed to providing what it can, even if it has to be purchased from other sources.[96] In April 2019, the Department of the Treasury moved to shut down that flow (and Cuban support for Maduro) by sanctioning 44 vessels and six shipping companies involved in transporting Venezuelan oil, including five companies that have transported Venezuelan oil to Cuba (see "U.S. Sanctions on Venezuela").[97]

Russia has remained a strong ally of the Maduro government. [98] Russia's trade relations with Venezuela are currently not significant. However, Venezuela was a major market for Russian arms between 2006 and 2013, with an estimated $4 billion in sales (mostly on credit).[99] Venezuela reportedly had more than 5,000 Russian-made surface-to-air missiles as of 2017, raising concern about the potential for them being stolen or sold to criminal or terrorist groups.[100] Russian state oil company Rosneft has also

---

[94] At that time, the report said that the number of Cuban military and intelligence advisors in Venezuela may have ranged from hundreds to thousands, coordinated by Cuba's military attaché in Venezuela. Ted Piccone and Harold Trinkunas, "The Cuba-Venezuela Alliance: The Beginning of the End?," Brookings Institution, June 2014.

[95] Venezuela Investigative Unit, "The Eight Criminal Armies Supporting Venezuela's Maduro Administration," Insight Crime, January 22, 2018; The White House, "President Donald J. Trump Stands for Democracy in Venezuela," May 1, 2019.

[96] "Venezuela Continues Shipping Oil to Cuba," *Oilprice.com*, April 12, 2019.

[97] U.S. Department of the Treasury, "Treasury Increases Pressure on Cuba to End Support to Maduro by Imposing Further Oil Sector Sanctions," April 12, 2019. For related information, see CRS Insight IN11120, *Cuba: Trump Administration Expands Sanctions*, by Mark P. Sullivan.

[98] See Vladimir Rouvinski, *Russian-Venezuelan Relations at a Crossroads*, Woodrow Wilson Center, February 5, 2019.

[99] Anthony Faiola and Karen DeYoung, "In Venezuela, Russia Pockets Key Energy Assets in Exchange for Cash Bailouts," *Washington Post*, December 24, 2018.

[100] Girish Gupta, "Exclusive: Venezuela Holds 5,000 Russian Surface-to-air MANPADS Missiles," Reuters, May 22, 2017.

invested billions of dollars in Venezuela.[101] Russia's 2017 decision to allow Venezuela to restructure $3 billion in debt provided much-needed financial relief to the Maduro government.[102]

Russia has called for the political crisis in Venezuela to be resolved peacefully and without outside interference. It reportedly views support for Maduro as a way to simultaneously oppose U.S. efforts at "regime change" and maintain influence in a region close to the United States.[103] In December 2018, Maduro visited Russia after which news reports suggested that Rosneft had lent PdVSA $6.5 billion, partly as a prepayment for crude oil.[104] That month, Russia also sent two nuclear-capable bombers to Venezuela to conduct joint exercises (the third such deployment since 2008).[105] The Russian government congratulated Maduro on his inauguration and has blocked U.N. Security Council action against Maduro. In March 2019, Russia sent an estimated 100 military personnel and associated equipment to Venezuela, a move the United States denounced.[106]

While Russia has both economic and geo-strategic interests in Venezuela and opposes what it views as U.S.-backed efforts at "regime change," China's interest in Venezuela is primarily economic, and the Chinese government has reportedly been in contact with the Guaidó government.[107] From 2007 through 2016, China provided some $62.2 billion

---

101 Christian Lowe and Rinat Sagdiev, "How Russia Sank Billions of Dollars into Venezuelan Quicksand," Reuters, March 14, 2019.

102 Rachel Ansley, "Debt Default Pushes Venezuela Further into Russian Orbit," Atlantic Council, November 22, 2017.

103 Frederick Kempe, "Russia's Venezuela Challenge," Atlantic Council, April 7, 2019; Emily Talkin, "Why Is Russia Clashing with the United States over Venezuela?" *Washington Post*, May 1, 2019.

104 Mery Mogollan and Daniel Rodriguez, "Maduro inks deal with Russia to boost Venezuela oil output 1 million b/d," *S&P Global Platts*, December 6, 2018.

105 Tom Phillips, "Venezuela Welcomes Russian Bombers in Show of Support for Maduro," *The Guardian*, December 10, 2018.

106 The White House, "Statement by National Security Advisor Ambassador John Bolton on Venezuela," March 29, 2019.

107 Stephen Kaplan and Michael Penfold, "China-Venezuela Economic Relations: Hedging Venezuelan Bets with Chinese Characteristics," Woodrow Wilson Center, February 20, 2019; Kejal Vyus, "China Talks with Venezuela Opposition to Protect Investments," *Wall Street Journal*, February 12, 2019.

in financing to Venezuela.[108] The funds have mostly been for infrastructure and other economic development projects, but China has also provided loans for military equipment to be repaid by oil deliveries.[109] Although the Chinese government has been patient when Venezuela has fallen behind on its oil deliveries, Beijing reportedly stopped providing new loans to Venezuela in fall 2016.[110] As previously stated, China, unlike Russia, has signaled a willingness to support dialogue efforts that could result in the convening of new presidential elections in Venezuela.[111]

Dialogue Facilitated by Norway

On May 29, 2019, the Norwegian government, which has remained neutral in the Venezuela crisis, announced that both Maduro and Guaidó representatives had demonstrated a "willingness" to continue trying to reach a political agreement to respond to the country's multifaceted crisis.[112] At least two rounds of exploratory talks have been held that have reportedly focused on the convening of new presidential elections with a new electoral authority.[113] Guaidó's supporters argue that Maduro should step down and allow a transitional government to organize those elections, a proposal he has thus far refused.[114] Other issues reportedly include political prisoners, the degree to which *chavistas* can participate in a transition government, humanitarian aid, disarming the *colectivos*, and how

---

[108] Kevin P. Gallagher and Margaret Myers, "China-Latin America Finance Database," Inter-American Dialogue, 2016, http://www.thedialogue.org/map_list/.

[109] R. Evan Ellis, Testimony to the U.S. Congress, House Committee on Foreign Affairs, Subcommittee on the Western Hemisphere, *The Influence of Extra-Hemispheric Actors on the Crisis in Venezuela*, 115th Cong., 1st sess., September 13, 2017.

[110] Marianna Parraga and Brian Ellsworth, "Venezuela Falls Behind on Oil-for-Loan Deals with China, Russia," Reuters, February 10, 2017; David Dollar, *Chinese Investment in Latin America*, Brookings Institution, January 2017.

[111] Ministry of Foreign Affairs of the People's Republic of China, Foreign Ministry Spokesperson Geng Shuang's regular press conference on May 8, 2019," May 8, 2019.

[112] Norwegian Ministry of Foreign Affairs, "Announcement About Venezuela," May 29, 2019.

[113] Sabrina Martín, "They Met Face to Face: Behind the Scenes at the Venezuelan Dialogues in Norway," *PanAmerican Post*, May 29, 2019.

[114] Terje and Berwick, "Venezuela's Guaidó Vows Protests."

to balance the need for justice for past abuses with a potential amnesty law for those who back a political transition.[115]

U.S. Policy

The United States has historically had close relations with Venezuela, a major U.S. foreign oil supplier, but friction in relations increased under the Chávez government and has intensified under the Maduro government. For more than a decade, U.S. policymakers have had concerns about the deterioration of human rights and democratic conditions in Venezuela and the lack of bilateral cooperation on counternarcotics and counterterrorism efforts. During this time, Congress has provided funding to support democratic civil society in Venezuela. As Maduro has become increasingly authoritarian, the Obama and Trump Administrations have increasingly turned to sanctions, first targeted on specific officials and then aimed at broader portions of the Venezuelan economy, to encourage a return to democracy in the country. More recently, U.S. humanitarian assistance has supported Venezuelans who have fled to neighboring countries.

Since January 2019, the United States has coordinated its efforts with Guaidó. On January 10, 2019, the U.S. State Department condemned Maduro's "illegitimate usurpation of power." President Trump recognized Guaidó as interim president of Venezuela on January 23 and has encouraged other governments to do the same. The Administration has imposed targeted sanctions (visa bans and sanctions blocking access to assets) on additional Maduro officials and their families, blocked the Maduro government's access to revenue from Venezuela's state oil company, and employed broader sanctions aimed at choking off other sources of revenue. In response to the interim government's request, the Administration has pre-positioned emergency assistance for Venezuelans on the country's borders. President Trump and other Administration officials have repeatedly asserted that "all options are on the table" to

[115] Martín, "They Met Face to Face."

address the Venezuela situation, including the use of U.S. military force.[116] Most analysts agree that a U.S. military intervention is unlikely. Some Members of Congress have expressed serious concerns about any U.S. military action in Venezuela, and such a policy is opposed by virtually all U.S. allies in the region.[117]

U.S. Sanctions on Venezuela[118]

The United States has increasingly employed sanctions as a policy tool in response to activities of the Venezuelan government and Venezuelan individuals. As the political and economic crisis in Venezuela has deepened, the Trump Administration has significantly expanded sanctions on Venezuela, relying on both existing sanctions and new executive orders. In particular, recent U.S. sanctions on Venezuela have targeted specific Venezuelan officials, the Maduro government's access to the U.S. financial system, and Venezuela's oil and gold sectors, which are key sectors and sources of financing for the Maduro government. The Trump Administration has cited a number of serious concerns about the Maduro government in expanding sanctions including, among others, human rights abuses, usurpation of power from the democratically elected National Assembly, rampant public corruption, degradation of Venezuela's infrastructure and environment through economic mismanagement and confiscatory mining and industrial practices, and its role in creating a regional migration crisis by neglecting the needs of the Venezuelan people.[119]

---

[116] The White House, "Remarks by President Trump to the Venezuelan American Community," February 18, 2019; Secretary of State Mike Pompeo, interview with Maria Bartiromo, *Mornings with Maria*, May 1, 2019.

[117] U.S. Congress, House Committee on Foreign Affairs, *Hearing on H.R. 1004, Prohibiting Unauthorized Military Action in Venezuela Act*, 116th Cong., 1st sess., March 13, 2019.

[118] This draws from: CRS In Focus IF10715, *Venezuela: Overview of U.S. Sanctions*, by Mark P. Sullivan.

[119] For example, see E.O. 13808, "Imposing Additional Sanctions with Respect to the Situation in Venezuela," 82 *Federal Register* 41155-41156, August 24, 2017; E.O. 13850, "Blocking Property of Additional Persons Contributing to the Situation in Venezuela," 83 *Federal Register* 55243-55245, November 1, 2018.

Targeted Sanctions on Venezuelan Officials

The Trump Administration has used existing authorities to sanction a number of Venezuelan officials. In particular, the Trump Administration has sanctioned 75 Venezuelan government and military officials pursuant to E.O. 13692, issued by President Obama in March 2015 to implement the Venezuela Defense of Human Rights and Civil Society Act of 2014 (P.L. 113- 278).[120] Among its provisions, the law requires the President to impose sanctions (asset blocking and visa restrictions) against those whom the President determines are responsible for significant acts of violence or serious human rights abuses associated with protests in February 2014 or, more broadly, against anyone who has directed or ordered the arrest or prosecution of a person primarily because of the person's legitimate exercise of freedom of expression or assembly. In 2016, Congress extended the 2014 act through 2019 in P.L. 114-194.

Under the Obama Administration, the Department of the Treasury froze the assets of seven Venezuelans—six members of Venezuela's security forces and a prosecutor involved in repressing antigovernment protesters—pursuant to E.O. 13692. Under the Trump Administration, the Department of the Treasury designated for sanctions an additional 75 Venezuelan government and military officials. These officials include Maduro and his wife, Cecilia Flores; Executive Vice President Delcy Rodriguez; PSUV First Vice President Diosdado Cabello; eight Supreme Court members; the leaders of Venezuela's army, national guard, and national police; four state governors; the director of the Central Bank of Venezuela; and the foreign minister.

The Trump Administration has also blocked the assets and travel of two high-ranking officials in Venezuela, including then-Vice President Tareck el Aissami and Pedro Luis Martin (a former senior intelligence official) pursuant to Foreign Narcotics Kingpin Designation Act (P.L. 106-120, Title VIII; 21 U.S.C. 1901 *et seq.*). Including those individuals, the Department of the Treasury has now imposed economic sanctions on at

---

[120] E.O. 13692, "Blocking Property and Suspending Entry of Certain Persons Contributing to the Situation in Venezuela," 80 *Federal Register* 12747-12751, March 8, 2015.

least 22 individuals with connections to Venezuela and 27 companies by designating them as Specially Designated Narcotics Traffickers pursuant to the Kingpin Act.

For a discussion on terrorism-related sanctions related to Venezuela, see "Terrorism" section below.

Sanctions Restricting Venezuela's Access to U.S. Financial Markets

Starting in 2017, President Trump has sought to cut off U.S. sources of financing for the Venezuelan government. In August 2017, President Trump issued E.O. 13808 to prohibit access to U.S. financial markets by the Venezuelan government, including PdVSA, with certain exceptions to minimize the impact on the Venezuelan people and U.S. economic interests.[121] The sanctions restrict the Venezuelan government's access to U.S. debt and equity markets, traditionally an important source of capital for the government and PdVSA.

In 2018, the Trump Administration issued two additional executive orders to further tighten Venezuela's access to U.S. currency. E.O. 13827, issued in March 2018, prohibits U.S. investors from purchasing or transacting in Venezuela's new digital currency, the petro. This is designed to help the government raise funds and circumvent U.S. sanctions.[122] E.O. 13835, issued in May 2018, prohibits U.S. investors from buying debt or accounts receivable with the Venezuelan government, including PdVSA. This is devised to close off an "avenue for corruption" used by Venezuelan government officials to enrich themselves.[123]

Sanctions on Venezuela's Oil and Gold Sectors

Beginning in late 2018, the Trump Administration has rolled out a series of new sanctions focused on Venezuela's oil, gold, financial,

---

[121] E.O. 13808, "Imposing Additional Sanctions with Respect to the Situation in Venezuela," 82 *Federal Register* 41155-41156, August 24, 2017.

[122] E.O. 13827, "Taking Additional Steps to Address the Situation in Venezuela," 83 *Federal Register* 12469-12470, March 19, 2018.

[123] E.O. 13835, "Prohibiting Certain Additional Transactions with Respect to Venezuela," 83 *Federal Register* 24001- 24002, May 21, 2018; Nicholas Casey and Julie Hirschfeld Davis, "As Trump Adds Sanctions on Venezuela, Its Neighbors Reject Election Result," *New York Times*, May 21, 2018.

defense, and security sectors. In November 2018, President Trump issued E.O. 13850, which blocks the assets of, and prohibits transactions with, those entities operating in Venezuela's gold sector (or any other sector of the economy as determined in the future by the Treasury Secretary) or responsible for or complicit in transactions involving deceptive practices or corruption in the Venezuelan government.[124] Three subsequent determinations issued by Secretary of the Treasury Mnuchin expanded the reach of E.O. 13850 to include the other sectors.[125]

Over subsequent months, a number of individuals and entities have been sanctioned pursuant to E.O. 13850, including five individuals involved in a corruption scheme involving Venezuela's currency exchange practices, PdVSA (with general licenses issued by Office of Foreign Assets Control [OFAC] to allow certain transactions and activities related to PdVSA and its subsidiaries), the Moscow-based Evrofinance Mosnarbank (jointly owned by Russia and Venezuela) for helping PdVSA funnel its cash flow from oil sales, Venezuela's state-owned gold sector company—Minerven—for using illicit gold operations to help Maduro financially, state-affiliated Venezuelan Economic and Social Development Bank and five of its subsidiaries that the Maduro government uses to move money outside of Venezuela, 44 vessels and six shipping companies involved in transporting Venezuelan oil, and Venezuela's central bank in order to cut off its access to U.S. currency and limit its ability to conduct international financial transactions.

Debate over the Efficacy of U.S. Sanctions on Venezuela

The Trump Administration has significantly ratcheted up sanctions on Venezuela, hitting key targets including Maduro himself, the state-owned oil and gold companies, and the Venezuelan central bank, among others. The Administration has also started targeting foreign firms doing business with Venezuela ("secondary sanctions") in its sanctions on foreign

---

[124] E.O. 13850, "Blocking Property of Additional Persons Contributing to the Situation in Venezuela," 83 *Federal Register* 55243-55245, November 1, 2018.

[125] Department of the Treasury determinations issued pursuant to Section 1(a)(1) of E.O. 13850 of January 28, 2019 (relating to oil), March 22, 2019 (relating to the financial sector), and May 9, 2019 (relating to the defense and security sector).

shipping companies transporting Venezuelan oil. At the same time, the Administration is starting to demonstrate some flexibility in its sanctions policy with the lifting of sanctions against the former head of Venezuela's intelligence service, General Manuel Cristopher Figuera, in May 2019 after he broke ranks with Maduro. To date, the Administration to date has stopped short of more aggressive sanctions, including on companies in other countries that purchase Venezuelan oil, although such sanctions have reportedly been considered.[126]

In terms of economic effects of the sanctions, the results are mixed. The sanctions have contributed to the Maduro government's financial problems, including default on most of its loans and inability to raise new financing.[127] However, the Maduro government has, to some extent, been able to work with Chinese and Russian governments to help fill financing gaps and continue to sell gold reserves to raise funds despite U.S. sanctions.[128] Some analysts are also concerned that the stronger sanctions on PdVSA are further exacerbating Venezuela's difficult humanitarian crisis, already marked by shortages of food and medicines and mass migration, by limiting the country's key source of revenue.[129] This is a particular concern in the event that Maduro remains in power over an extended period of time.

In terms of the sanctions' political effects, the imposition of targeted sanctions on individuals in the Maduro government has not encouraged many of those who were not yet sanctioned to abandon Maduro or changed the behavior of the sanctioned individuals. Instead, some officials (such as former vice president Tareck el Aissami and interior minister Nester Reverol) received promotions after being designated for sanctions by the United States, making their futures more dependent on remaining loyal to

---

[126] Steve Holland and Roberta Rampton, "Exclusive: Trump Eyeing Stepped-Up Venezuela Sanctions for Foreign Companies—Bolton," Reuters, March 29, 2019.

[127] Dany Bahar, "Chavismo Is the Worst of All Sanctions: The Evidence Behind the Humanitarian Catastrophe in Venezuela," Brookings Institution, May 22, 2019.

[128] Patricia Laya, "Venezuela Sells $570 Million From Gold Reserve Despite Sanctions," *Bloomberg*, May 17, 2019.

[129] Kurmanaev and Krauss, "U.S. Sanctions Are Aimed at Venezuela's Oil;" Jim Wyss, "As U.S. Sanctions Against Venezuela Mount, What's the Human Toll?," *Miami Herald*, March 12, 2019.

Maduro. While some have praised the Administration for removing the sanctions on General Figuera after he backed Guaidó, others have questioned how willing or able the U.S. government will be to lift sanctions on others, particularly in cases involving the Kingpin Act or for those who face U.S. criminal indictments.[130] Observers have urged the Administration to coordinate the imposition or removal of future sanctions, including travel restrictions, with allies such as the EU and Canada, which have enacted similar sanctions.[131] Broader U.S. sanctions adopted since 2017 have yet to compel Maduro to leave office despite the country's increasingly dire economic situation. They have also provided a foil on which Maduro has blamed the country's economic problems.

Energy Sector Concerns and U.S. Economic Sanctions[132]

Venezuela's petroleum sector has experienced a general decline since the beginning of 2011, when crude oil production averaged approximately 2.5 million barrels per day (bpd) for the year. Prior to financial sanctions being imposed on PdVSA in August 2017 (E.O. 13808), oil production had already declined to approximately 1.8 million bpd.

Sanctions included in E.O. 13808 place limits on the maturity of new debt that U.S. entities can extend to PdVSA to 90 days. This limitation made it more difficult for PdVSA to pay for oil-related services and acquire certain oil production equipment. PdVSA has dealt with its fiscal problems by delaying payments and paying service providers with promissory notes in lieu of payments.[133] There are concerns that delayed payments and promissory notes would count as new credit and, if their maturity exceeds 90 days, would violate sanctions. These payment issues have contributed to the slowdown in oil production, although they have not

---

130 Reuters, "Relief from U.S. Sanctions Will Not Come Easily for Some Venezuelans," May 30, 2019.

131 Samantha Sultoon, "Spotlight: Next Steps with Venezuela," Atlantic Council, April 25, 2019.

132 This section was authored by Philip Brown, Specialist in Energy Policy.

133 Risa Grais-Targow et al., "Venezuela/Oil," Eurasia Group, November 21, 2017.

halted it.[134] The financial sanctions also prevent PdVSA from receiving cash distributions from its U.S.-based Citgo refining subsidiary. While it is difficult to accurately quantify the impact of E.O. 13808 financial sanctions, data in Figure 4 suggest that Venezuela's oil production decline accelerated following the imposition of financial access restrictions on PdVSA.

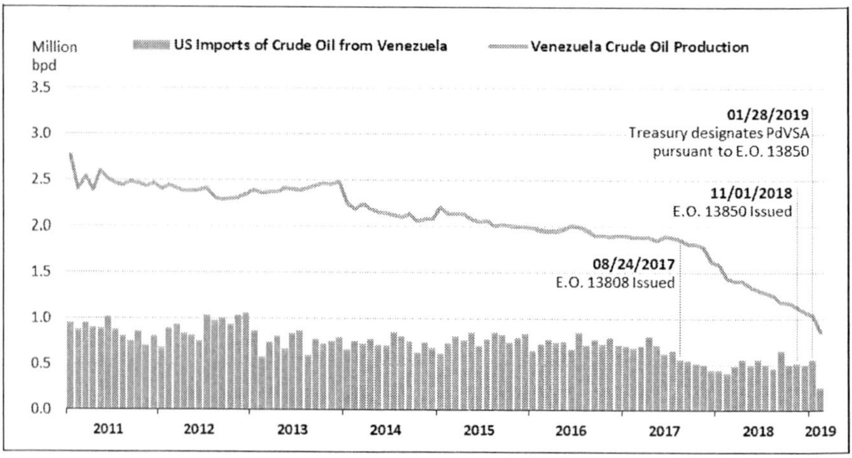

Source: Venezuela crude oil production data from Energy Intelligence via Bloomberg L.P. U.S. import data from the Energy Information Administration.

Figure 4. Venezuela Crude Oil Production and U.S. Imports. January 2011-February 2019.

As previously described, the Secretary of the Treasury determined on January 28, 2019 that persons (e.g., individuals and companies) operating in Venezuela's oil sector are subject to sanctions—pursuant to E.O. 13850—the goal being to apply economic pressure on the Maduro government. Subsequently, Treasury's OFAC added PdVSA—including all entities in which PdVSA has a 50% or more ownership position—to its Specifically Designated Nationals list.

[134] Grais-Targow et al., "Venezuela/Oil."

This designation blocks PdVSA's U.S. assets and prohibits U.S. persons from dealing with the company. These sanctions will affect several areas in which U.S. companies have business interests (e.g., debt/financial transactions and oil field services) and will effectively terminate U.S.-Venezuela petroleum (crude oil and petroleum products) trade. This is significant for Venezuela, as the United States was the primary cash buyer for exported crude oil.

Petroleum trade between the United States and Venezuela has been bilateral but has historically been dominated by Venezuela crude oil exports to the United States. Similar to oil production trends in Venezuela, U.S. oil imports of Venezuelan crude oil have been steadily declining since at least 2011 (see Figure 4) when import volumes were approximately 1 million bpd during certain months. When sanctions were imposed on PdVSA in January 2019, U.S. oil imports of Venezuelan crude oil were approximately 500,000 bpd. General licenses issued by OFAC allowed for U.S. entities to continue purchasing Venezuelan crude oil until April 28, 2019.[135] However, payments made to PdVSA for imported petroleum during the wind-down period had to be made into a U.S.-based blocked account. This element of the petroleum trade sanctions motivated PdVSA to immediately seek alternative cash buyers.[136] U.S. imports declined by 50% to approximately 250,000 bpd in the first month following the PdVSA designation.

Venezuela's oil production since January 2019 has been further challenged by access to petroleum blending components—referred to as diluents—and ongoing power outages that have affected oil production operations.[137]

---

[135] General license 12 provides this wind-down period for importing petroleum from Venezuela. A comprehensive list of OFAC general licenses is available at https://www.treasury.gov/resource-center/sanctions/programs/pages/ venezuela.aspx.

[136] Ship tracking information from Bloomberg L.P. (subscription required) indicates that Venezuela's crude exports to India and China increased 49% and 34%, respectively, between January 2019 and February 2019.

[137] Energy Information Administration (EIA), "Venezuelan Crude Oil Production Falls to Lowest Level Since January 2003," *Today in Energy*, May 20, 2019.

PdVSA sanctions imposed in January 2019 prohibited U.S. diluent exports to Venezuela immediately. There was no wind-down period for this trade element. The characteristics of a large portion of Venezuela's crude oil—generally a heavy, low-viscosity, high-sulfur crude oil type—requires blending this crude oil with lighter petroleum materials in order to achieve a viscosity sufficient for transportation and processing. Typical diluents used for crude oil blending include naphtha and condensate, both light petroleum liquids that Venezuela has been importing from the United States. There are other non-U.S. diluent sources, and PdVSA will have to source diluent materials from alternative suppliers. However, it may take some time to secure alternative suppliers, and this constraint causes additional difficulty for PdVSA to continue its oil production operations. Furthermore, power outages in Venezuela have been reported since the January sanctions took effect.[138] Limited power access is also creating difficulties for PdVSA oil production activities as pumping and processing equipment require power to operate.[139]

Direct impacts to U.S. oil refineries are somewhat difficult to assess due to the wind-down period provided by general licenses, opportunities to source similar crude oil types from other nonsanctioned countries, and refinery maintenance activities during the first half of 2019 that limited demand for oil import volumes. Mexico, Colombia, Canada, and Iraq are examples of countries that export crude oil with similar quality characteristics to that of Venezuela's heavy crude oil. However, movements in oil price benchmarks since January indicate that the differential between light sweet (low sulfur) and heavy sour (high sulfur) crude oil has narrowed. This price behavior can affect profit margins for refineries optimally configured to process heavy crude oil.[140]

---

138 EIA, "Venezuelan Crude Oil Production."

139 EIA, "Venezuelan Crude Oil Production."

140 The price differential between Louisiana Light Sweet (LLS) and Mars—a U.S.-produced heavy crude oil—ranged from approximately $2.50 to $6.00 per barrel between 2016 and 2018. Mars was less than LLS, which is to be expected given the better quality characteristics of LLS. In March 2019, following the January 2019 sanctions, the price differential turned negative for one day by $0.90 per barrel. Mars, a lower quality crude oil, was more expensive than LLS. The price differential has since recovered to approximately $2.50 per barrel, the lower end of the 2016-2018 range.

Citgo's Uncertain Future: Multiple Creditors, Legal Claims, and New Management

PdVSA's U.S.-based Citgo subsidiary owns and operates oil refineries in Texas, Louisiana, and Illinois. The company also owns 48 petroleum product terminals and a pipeline network that delivers these products to various customers. Citgo is arguably one of the most valuable assets in the PdVSA portfolio. PdVSA has looked to leverage that value to support its borrowing activities, as described below. All of Citgo's ownership has been used as collateral for two separate debt issuances (bonds and a loan to Rosneft, as described below). Additionally, companies that have sued Venezuela—such as Crytallex, discussed below—for expropriation actions have been awarded legal judgments that include taking possession of and liquidating Citgo assets.[141] Furthermore, the U.S.- recognized Guaidó government appointed a new board of directors in February 2019 to manage Citgo, thereby removing Maduro's PdVSA from management decision-making of the U.S.-based company. Some of the financial and legal challenges facing Citgo are described below. PdVSA 2020 bonds (50.1% of Citgo as collateral): PdVSA issued bonds totaling approximately $3.4 billion in October 2016 to various creditors.[142] A majority ownership stake in Citgo was pledged as collateral for the bonds, and the creditors could have legal claim to the company should the bonds enter into default.[143] An interest payment for the bonds came due in April 2019.[144] The payment was missed, but there is a 30-day grace period for the missed payment.[145] The Maduro-controlled PdVSA—the original bond issuer—is prevented from making bond payments due to U.S. sanctions on PdVSA. However, a PdVSA board appointed by Guaidó voted to pay the $71 million interest payment, thus preserving Citgo ownership in the event that the opposition takes control of the Venezuela government. An additional $913 million payment is due in October 2019.

Rosneft loan to PdVSA (49.9% of Citgo as collateral): PdVSA pledged 49.9% of Citgo ownership as collateral for a $1.5 billion loan from Russian oil company Rosneft that was issued in December 2016.[146] The potential for a Russian company to own a significant portion of the U.S.-based Citgo operation has been noted by Members of Congress as a potential national security concern.[147] In the 116th Congress, S. 1025 has been introduced and includes Section 609, which express concerns about Citgo ownership pledged as collateral for the Rosneft loan and requires the President to prevent Rosneft from taking control of U.S. energy infrastructure.

---

141 Other arbitration awards have been made to companies such as Conoco for their expropriation in 2007 and Canada's Rusoro Mining, whose gold projects were seized by the government of Venezuela in 2011. However, the Crystallex award and judgment allows the company to seize and liquidate Citgo shares as a means of collecting the award.

142 Bloomberg L.P. (subscription required), Bond description for PdVSA 8 1/2 10/27/20 bonds, accessed May 21, 2019.

143 PdVSA bond prospectus, October 2016.

144 Reuters, "Investors Await Payment Decision as PDVSA 2020 Coupon Looms," April 24, 2019.

145 Reuters, "Investors Await Payment Decision."

146 Reuters, "Venezuela's PDVSA Uses 49.9 Percent Citgo Stake as Loan Collateral," December 23, 2016.

147 Members of the Senate Finance Committee sent a letter to the Secretary of the Treasury and the Secretary of State in June 2017 expressing concerns about national security implications of the potential for Russia to control a large portion of Citgo. A copy of the letter is available at https://www.finance.senate.gov/imo/.../061917%20Rosneft%20Citgo%20 Letter1.pdf.

Crystallex legal judgment ($1.2 billion judgment against PDV Holding, Citgo's parent company): In 2018, a U.S. district court ruled that Canada-based Crystallex could take shares of PDV Holding, the U.S.-based parent company of Citgo, as a means of collecting a $1.2 billion arbitration award made to Crystallex.[148] The award stems from Venezuela's seizure of the Canadian miners' assets in 2011. Crystallex has indicated its intent to take ownership of Citgo shares and sell them for cash as a means of collecting the arbitration award.

U.S. Humanitarian and Related Assistance[149]

The U.S. government is providing humanitarian and emergency food assistance and helping to coordinate and support regional response efforts. As of April 10, 2019 (latest data available), the United States has committed to providing more than $213.3 million since FY2017 for Venezuelans who have fled to other countries and for the communities hosting Venezuelan refugees and migrants, including nearly $130 million to Colombia.[150] (Humanitarian funding is drawn primarily from the global humanitarian accounts in annual Department of State/Foreign Operations appropriations acts.) From October through December 2018, the U.S. Navy hospital ship USNS *Comfort* was on an 11-week medical support deployment to work with government partners in Ecuador, Peru, Colombia, and Honduras, in part to assist with arrivals from Venezuela. The *Comfort* is scheduled to deploy for another five-month deployment in June.[151]

In Colombia, the U.S. response aims to help the Venezuelan arrivals as well as the local Colombian communities that are hosting them. In addition to humanitarian assistance, the United States is providing $37 million in bilateral assistance to support medium and longer-term efforts by Colombia to respond to the Venezuelan arrivals.

---

148 Oil Daily, "Judge Paves Way for Canadian Firm to Seize Citgo Assets," August 13, 2018.

149 This is drawn from CRS In Focus IF11029, *The Venezuela Regional Migration Crisis*, by Rhoda Margesson and Clare Ribando Seelke.

150 U.S. Department of State, "United States Provides Additional Humanitarian Aid to Venezuelans Who Have Fled the Country," April 10, 2019. See also USAID, *Venezuela Regional Crisis*.

151 Ben Werner, "USNS Comfort Heading to Colombia to Treat Venezuelan Refugees," *USNI News*, May 7, 2019.

In February 2019, the United States responded to Guaidó's request for immediate international humanitarian assistance. To date, along with other partners working in the Western Hemisphere, the U.S. government has pre-positioned nearly 546 metric tons of assistance in Colombia, Brazil, and Curacao to help tens of thousands of Venezuelans, including 208 metric tons in the border city of Cúcuta alone. The aid includes food, medical supplies, hygiene kits, nutrition products, and water purification and storage units. According to the Colombian government, some of that assistance is now going to be distributed to Venezuelans and communities sheltering them in Colombia.[152]

U.S. Democracy Assistance

For more than a decade, the United States has provided democracy-related assistance to Venezuelan civil society through the U.S. Agency for International Development (USAID) and the National Endowment for Democracy (NED).

From 2002 through 2010, USAID supported small-grant and technical assistance activities through its Office of Transition Initiatives (OTI) to provide assistance monitoring democratic stability and strengthening the county's democratic institutions. At the end of 2010, USAID's support for such activities in Venezuela was transferred from OTI to USAID's Latin America and Caribbean Bureau. U.S. democracy and human rights assistance to Venezuela amounted to $15 million to civil society groups in Venezuela in FY2018 and $17.5 million in FY2019.

The Administration's proposed FY2020 budget asks for $9 million in democracy aid and authority to transfer up to $500 million to support a transition or respond to a crisis in Venezuela. On May 20, 2019, the House Appropriations Committee reported its version of the FY2020 Department of State and Foreign Operations Appropriations Act (H.R. 2389), which

---

[152] Associated Press, "US Emergency Aid for Venezuela to Be Distributed in Colombia," May 30, 2019.

would provide $17.5 million in democracy and human rights aid to Venezuela.

NED has funded democracy projects in Venezuela since 1992. U.S. funding for NED is provided in the annual State Department and Foreign Operations appropriations measure, but country allocations for NED are not specified in the legislation. In 2018, NED funded 40 projects in Venezuela totaling roughly $2.0 million as compared to $2.6 million in 2017.[153]

Organized Crime-Related Issues

Venezuela has among the highest crime victimization and homicide rates in Latin America and the Caribbean, the region with the highest homicide rates in the world.[154] According to the Venezuelan Violence Observatory, the homicide rate in Venezuela declined in 2018 (81.4 homicides per 100,000 people) as compared to a rate of 89.1 per 100,000 people in 2017, with part of that decline attributed to migration that has reduced the population.[155] The impunity rate for homicide in Venezuela is roughly 92%.[156] Although many homicides have been committed by criminal groups, extrajudicial killings by security forces and allied armed civilian militias (*colectivos*) have also been rising.[157] In September 2018, Amnesty International published a report describing how security forces have adopted militarized approaches to public security that have resulted in numerous human rights abuses, including extrajudicial killings.[158]

---

[153] National Endowment for Democracy, "Venezuela 2018," https://www.ned.org/region/latin-america-and-caribbean/ venezuela-2018/.

[154] Laura Jaitman, ed., *The Costs of Crime and Violence in Latin America: New Evidence and Insights from Latin America and the Caribbean*, Inter-American Development Bank, 2017.

[155] Reuters, "Venezuela Murder Rate Dips, Partly due to Migration: Monitoring Group," December 27, 2018.

[156] Christopher Woody, "Venezuela Admits Homicides Soared to 60 a day in 2016, Making it one of the Most Violent Countries in the World," *Business Insider,* April 2017.

[157] David Smilde, "Crime and Revolution in Venezuela," *NACLA Report on the Americas*, 2017, vol. 49, no. 3; Amnesty International, September 2018.

[158] Amnesty International, *This Is No Way to Live: Public Security and Right to Life in Venezuela*, September 20, 2018.

A May 2018 report by Insight Crime identified more than 120 high-level Venezuelan officials who have engaged in criminal activity, which has blurred the lines between crime groups and the state.[159] Many of those officials have allegedly engaged in drug trafficking (discussed below), but others have reportedly deputized illegal groups in the neighborhoods and prisons, run smuggling operations in border areas, and engaged in illegal gold mining.[160]

Counternarcotics

Venezuela's pervasive corruption and extensive 1,370-mile border with Colombia have made the country a major transit route for cocaine destined for the United States and an attractive environment for drug traffickers and other criminals to engage in money laundering. In 2005, Venezuela suspended its cooperation with the U.S. Drug Enforcement Administration (DEA) after alleging that DEA agents were spying on the government, charges U.S. officials dismissed as baseless. Prior to that time, the governments had negotiated an antidrug cooperation agreement (an addendum to a 1978 Bilateral Counternarcotics agreement) that would have enhanced information-sharing and antidrug cooperation. Venezuela has yet to approve that agreement.

Since 2005, Venezuela has been designated annually as a country that has failed to adhere to its international antidrug obligations, pursuant to international drug-control certification procedures in the Foreign Relations Authorization Act, FY2003 (P.L. 107-228). In September 2018, President Trump designated Venezuela as one of two countries not adhering to its antidrug obligations.[161] At the same time, President Trump waived economic sanctions that would have curtailed U.S. assistance for democracy programs.

The State Department reported in its 2019 *International Narcotics Control Strategy Report* (INCSR) that Venezuela was one of the preferred

---

159 Insight Crime, *Venezuela: A Mafia State?*, May 2018.

160 International Crisis Group, *Gold and Grief in Venezuela's Violent South.*

161 The White House, "Presidential Determination—Major Drug Transit or Major Illicit Drug Producing Countries for Fiscal Year 2018," September 11, 2018.

trafficking routes for the transit of illicit drugs out of South America, especially cocaine, because of the country's porous border with Colombia, economic crisis, weak judicial system, lacking international counternarcotics cooperation, and permissive and corrupt environment. The report notes the following:

- Cocaine is trafficked via aerial, terrestrial, and maritime routes, with most drug flights departing from Venezuelan states bordering Colombia and maritime trafficking that includes the use of large cargo containers, fishing vessels, and "go-fast" boats.
- The vast majority of drugs transiting Venezuela in 2018 were destined for the Caribbean, Central America, the United States, West Africa, and Europe. Colombian drug-trafficking organizations—including multiple criminal bands, the Revolutionary Armed Forces of Colombia (FARC), and the National Liberation Army (ELN)—facilitate drug transshipment through Venezuela. Mexican drug-trafficking organizations also operate in the country.
- Venezuelan officials reported conflicting statistics, with the attorney general asserting that the government had seized 20 metric tons of drugs between August 2017 and October 2018 and another official maintaining that seizures from January through September 2018 alone exceeded 22 metric tons.
- Venezuelan authorities "failed to make demonstrable efforts to combat illegal drug activity and prosecute corrupt officials or suspected drug traffickers, including those sanctioned by the United States government."[162]

In addition to State Department reports, a report by Insight Crime entitled *Drug Trafficking within the Venezuelan Regime: the Cartel of the Suns* describes in detail how the Venezuelan military, particularly the

[162] U.S. Department of State, Bureau of International Narcotics and Law Enforcement Affairs, 2019 INCSR, vol. 1, p. 271.

national guard, has been involved in the drug trade since 2002.[163] It names officials who have been sanctioned or accused of drug trafficking-related crimes, as well as others for whom there is significant evidence of their involvement in the drug trade. Insight Crime has also documented how the Cartel of the Suns has interacted with illegally armed groups and drug traffickers in Colombia, trafficked cocaine through the Dominican Republic and Honduras, and engaged in corruption with politicians and businesses in El Salvador.[164]

Recent cases in the United States also demonstrate the involvement of high-level Venezuelan officials or their relatives in international drug trafficking. President Maduro has either dismissed those cases or appointed the accused to Cabinet positions, where they will presumably be protected from extradition. Some observers have maintained that it may therefore be difficult to persuade officials to leave office through democratic means if, once out of power, they would likely face extradition and prosecution in the United States.[165]

In December 2017, two nephews of First Lady Cilia Flores—Franqui Francisco Flores de Freitas and Efraín Antonio Campo Flores—were sentenced to 18 years in a U.S. federal prison for conspiring to transport cocaine into the United States. The two nephews had been arrested in Haiti in November 2015 and convicted in the United States in November 2016.[166]

Money Laundering

In addition to drug trafficking, the 2019 INCSR discusses Venezuela's high level of vulnerability to money laundering and other financial crimes. According to the report, money laundering is widespread in the country and is evident in industries ranging from government currency exchanges

---

163 Insight Crime, *Drug Trafficking Within the Venezuelan Regime: The 'Cartel of the Suns,'* May 17, 2018.

164 Insight Crime, *Venezuela: A Mafia State?*

165 U.S. Congress, Senate Committee on Foreign Relations, *Venezuela: Options for U.S. Policy*, 115th Cong., 1st sess., March 2, 2017.

166 Laura Natalia Ávila, "Game Changers 2016: Venezuela's Cartel of the Suns Revealed," Insight Crime, January 3, 2017.

to banks to real estate to metal and oil. Venezuela's currency-control system requires individuals and firms to purchase hard currency from the government's currency commission at a fixed exchange rate, which has created incentives for trade-based money laundering.

Venezuela revised its laws against organized crime and terrorist financing in 2014 but excluded the government and state-owned industries from the scope of any investigations. The unit charged with investigating financial crimes has "limited operational capabilities," and there is a lack of political will in the judicial system to combat money laundering and corruption.[167] The 2019 INCSR concludes that Venezuela's "status as a drug transit country, combined with weak AML supervision and enforcement, lack of political will, limited bilateral cooperation, an unstable economy, and endemic corruption" make the country vulnerable to money laundering.[168] As an example, in mid-June 2018, a U.S. district judge sentenced the Florida owners of a construction equipment export company who had been found guilty of laundering and transferring $100 million from Venezuela to bank accounts in the United States and other countries.[169]

On September 20, 2017, the Department of the Treasury's Financial Crimes Enforcement Network advised U.S. financial institutions to report any suspicious financial transactions that could have a nexus with Venezuela.[170] The advisory urges U.S. institutions to exercise increased scrutiny over transactions that may involve lesser-known state-owned enterprises connected to the government. It also warns that recent sanctions against Venezuelan officials could "increase the likelihood that other non-designated Venezuelan senior political figures may seek to protect their assets."

---

[167] U.S. Department of State, 2019 INCSR, pp. 193-194.

[168] U.S. Department of State, 2019 INCSR, pp. 193-194.

[169] Brendan Pierson, "Florida Men Sentenced to Prison for Laundering Funds From Venezuela," Reuters, June 19, 2018.

[170] U.S. Department of the Treasury, Financial Crimes Enforcement Network, "Advisory on Widespread Public Corruption in Venezuela," September 20, 2017.

Illegal Mining

Although more than 95% of Venezuela's export revenue comes from oil and gas exports, gold mining, both licit and illicit, has accelerated as the country's economy has collapsed in the face of low global oil prices and an ongoing political crisis. According to the Global Initiative against Transnational Organized Crime, 91% of gold produced in Venezuela was mined illegally—the highest rate in Latin America, even prior to the current crisis.[171] Over the past three years, a boom in illegal mining in Venezuela has reportedly contributed to deforestation and environmental degradation in indigenous areas, clashes between rival criminal gangs and violence committed by those gangs against miners whom they extort, and an outbreak of malaria (a disease that had been eradicated).[172] According to numerous reports, the illegal mining industry also commits various human rights violations, reportedly including the forcible recruitment of child labor from the indigenous Yanomami tribe.[173]

Colombian Illegally Armed Groups Operating in Venezuela

Illegally armed groups are active on both sides of the Colombian-Venezuelan border. Venezuela's instability, weak institutions, and lawlessness have attracted the attention of drug trafficking and illegally armed groups from Colombia. The ELN, which is still engaged in armed conflict in Colombia, and its rival, the Popular Liberation Army (EPL), reportedly recruit Venezuelans to cultivate coca, the basic plant component of cocaine.[174] Colombia's ELN guerrillas have moved from seeking safe haven in Venezuela to taking control of illicit gold mining areas in some parts of the country. The Rastrojos, a criminal group that contains former

---

171 Global Initiative Against Transnational Organized Crime, *Organized Crime and Illegally Mined Gold in Latin America*, April 2016.

172 Bram Ebus, "Militarization and Mining a Dangerous Mix in Venezuelan Amazon, " Mongabay, December 7, 2017; Maria Isabel Sanchez, "Inside the Deadly World of Venezuela's Illegal Mines," AFP, March 19, 2017; Stephanie Nebehay, "Malarian on Rise in Crisis-Hit Venezuela, WHO Says," Reuters, April 24, 2018.

173 Global Initiative Against Transnational Organized Crime, *Organized Crime and Illegally Mined Gold in Latin America*; International Crisis Group, *Gold and Grief in Venezuela's Violent South*.

174 A May 2018 report by Insight Crime identified more than 120 high-level Venezuelan officials who have engaged in criminal activity. Insight Crime, *Venezuela: A Mafia State?*

paramilitaries, reportedly controls important gasoline smuggling routes between Venezuela and Colombia. Human trafficking and sexual exploitation of Venezuelan migrants is prevalent in Colombia and border regions straddling the countries. Many observers assert that dissident FARC guerrillas are using border areas to regroup.[175]

Violence among these groups and between the groups and the Venezuelan government has escalated, threatening security on both sides of the border. Conflict between the ELN and the EPL over control of the cocaine trade led to an August 2018 daytime shootout in a town on the Colombian side of the border in which eight people died.[176] Since early 2018, Freddy Bernal, an official on the U.S. Kingpin List who allegedly supplied arms to the FARC, has served as head of security in Táchira state bordering Colombia. After Bernal ordered an elite police unit to arrest members of the Rastrojos, the group attacked a Venezuelan military base in October 2018, killing three soldiers.[177] The ELN reportedly killed three Venezuelan national guardsmen in Amazonas state in November 2018.[178] As this violence has occurred, Colombia has also protested periodic crossings into its territory by Venezuelan troops.[179]

Terrorism

The Secretary of State has determined annually, since 2006, that Venezuela has not been "cooperating fully with United States antiterrorism efforts" pursuant to Section 40A of the Arms Export Control Act (AECA). Per the AECA, such a designation subjects Venezuela to a U.S. arms

---

[175] Fundación Ideas para la Paz, "Inseguridad y Violencia en las Fronteras, los Desafíos del Nuevo Gobierno," November 7, 2018.

[176] Ronna Rísquez and Victoria Dittmar, " ELN and EPL Conflict Intensifies at Colombia-Venezuela Border," Insight Crime, August 2, 2018.

[177] "Rastrojos Demonstrate Power with Attack on Venezuela Military Base," Insight Crime, November 5, 2018.

[178] "Mueren Tres GNB en Enfrentamientos con Grupos Irregulares Colombianos en Amazonas," *Crónica Uno*, November 4, 2018.

[179] Reuters, "Colombia Protests New Border Crossing by Venezuelan Troops," November 7, 2018.

embargo, which prohibits all U.S. commercial arms sales and retransfers to Venezuela. The most recent determination was made in May 2019.[180]

In 2008, the Department of the Treasury imposed sanctions (asset freezing and prohibitions on transactions) on two individuals and two travel agencies in Venezuela for providing financial support to Hezbollah, which the Department of State has designated a Foreign Terrorist Organization. The action was taken pursuant to E.O. 13224, aimed at impeding terrorist funding.

The State Department's most recent annual terrorism report, issued in September 2018, stated that "country's porous borders offered a permissive environment to known terrorist groups."[181] Unlike in years past, the report did not identify any specific terrorist groups or sympathizers present in the country.[182] This designation would trigger an array of sanctions, including aid restrictions, requirement for validated export licenses for dual-use items, and other financial restrictions.[183]

Critics caution there is a lack of evidence to conclude that the Venezuelan government has "repeatedly provided support for acts of international terrorism," as required by law.[184]

OUTLOOK

For years, analysts have debated how long Maduro can retain his grip on power amid a deepening economic and humanitarian crisis and how

---

180 U.S. Department of State. Public Notice 10781, "Determination and Certification of Countries Not Cooperating Fully with United States Antiterrorism Efforts," 84 *Federal Register* 24856, May 29, 2019.

181 U.S. Department of State, "Western Hemisphere Overview," in *Country Reports on Terrorism 2017*, September 2018.

182 The report covering 2016 stated that individuals linked to the FARC, the ELN, and Basque Fatherland and Liberty (a Basque terrorist organization), as well as Hezbollah supporters and sympathizers, were present in Venezuela. U.S. Department of State, "Western Hemisphere Overview," in *Country Reports on Terrorism 2016*, June 2017.

183 CRS Report R43835, *State Sponsors of Acts of International Terrorism—Legislative Parameters: In Brief*, by Dianne E. Rennack.

184 Amanda Erickson, "Venezuela Is a Tragedy, Not a Terrorist Threat," *Washington Post*, November 21, 2018.

best to help hasten a return to democracy in Venezuela. With his second term widely regarded as illegitimate within Venezuela and by large parts of the international community, Maduro faces increasing threats to his control over the country. Under the leadership of Juan Guaidó, the opposition has come closer to ousting Maduro than ever before. Maduro still controls the top echelon of the military, but arrests of high-level military officials and a few recent defections suggest there is considerable dissent within the forces. Although nascent efforts at dialogue have gotten underway in Norway, the situation in the country remains volatile, and the current stalemate appears untenable due to the country's rapidly deteriorating economic and humanitarian conditions.[185]

The Trump Administration has worked bilaterally and multilaterally to increase pressure on Maduro and support the interim government led by Guaidó. In the aftermath of a failed April 30 uprising, U.S. options to hasten a transition to democracy in Venezuela appear limited. Experts have suggested that maintaining concerted and coordinated international pressure is required to compel Maduro to leave office.[186] While some observers have urged the Administration to take even more aggressive measures despite the risk that they could contribute to increased unrest, others have maintained that support for a negotiated solution is the best course of action. In the absence of a political transition, the deteriorating humanitarian situation could increase pressure on the United States to reassess its sanctions on the Venezuelan economy.

The 116th Congress has closely followed developments in Venezuela, Trump Administration's policy responses, and international efforts to broker a solution to the crisis. Congress has held hearings on the political crisis in Venezuela and U.S. policy responses, the humanitarian crisis in Venezuela, the regional migration crisis that Venezuela's unrest has wrought throughout Latin America and the Caribbean, the influence of Russia and China in Venezuela, and the role of Congress in authorizing the possible use of U.S. military force in Venezuela. Congress is also considering a range of legislative initiatives to require, authorize, or

[185] Moises Rendón, "What Happens Next in Venezuela?," CSIS, May 10, 2019.
[186] Samantha Sultoon, "Spotlight: Next Steps with Venezuela," Atlantic Council, April 25, 2019.

constrain certain Administration actions regarding Venezuela (e.g., H.R. 920, H.R. 1477, S. 1025, H.R. 1004, and S.J.Res. 11). For more information on legislative initiatives on Venezuela in the 116th Congress, see Appendix A.

International perspectives—particularly from UNHCR, IOM, and the Lima Group—may influence oversight of the $213 million in U.S. humanitarian assistance allocated thus far to help support Venezuelans in the region. They may also inform decisions about the amounts and types of U.S. funds most needed to support international organizations and U.N. agencies working inside Venezuela both now and in the future. The Administration's proposed FY2020 budget requests $9 million in democracy aid through the Economic Support and Development Fund account and the authority to transfer up to $500 million to support a transition or respond to a crisis in Venezuela. Should Maduro leave office, the Administration proposes that such funds could support international election observers, increased humanitarian assistance inside Venezuela, and/or a potential IMF package.

APPENDIX A. LEGISLATIVE INITIATIVES IN THE 116TH CONGRESS

Enacted Legislation

P.L. 116-6 (H.J.Res. 31). Consolidated Appropriations Act, 2019. Introduced January 22, 2019. House passed (231-180) January 24; Senate passed, amended, by voice vote January 25. Conference report (H.Rept. 116-9) filed February 13, 2019. House approved conference (300- 128) February 14; Senate approved conference (83-16) February 14. Signed into law February 15, 2019.

The measure provides $17.5 million for democracy and rule of law programs for civil society groups in Venezuela. H.Rept. 116-9 requires a strategy of how U.S. agencies are supporting communities that are sheltering Venezuelans throughout Latin America and the Caribbean.

Select Legislative Initiatives

The Venezuela TPS Act of 2019, H.R. 549 (Soto), would allow certain Venezuelan nationals residing in the United States to qualify for Temporary Protected Status, which prevents their removal from the United States and allows them to obtain employment and travel authorization. Introduced January 15, 2019; amended and reported out of the House Judiciary Committee May 22, 2019.

The Humanitarian Assistance to the Venezuelan People Act of 2019, H.R. 854 (MurcarselPowell), would require a strategy within 180 days of its enactment from the Department of State and USAID on the delivery of humanitarian assistance within Venezuela and for Venezuelans throughout Latin America and the Caribbean and authorize up to $150 million in humanitarian assistance to be provided in FY2020 and in FY2021. Introduced January 29, 2019; amended and reported out of the House Foreign Affairs Committee March 14, 2019; approved by the House, as amended, March 25, 2019.

The Venezuela Arms Restriction Act, H.R. 920 (Shalala), would restrict the transfer of defense articles, defense services, and crime control articles to any element of the security forces of Venezuela under the authority of a government of Venezuela that is not recognized as the legitimate government of Venezuela by the government of the United States (i.e., the Maduro government). Introduced January 30, 2019; reported out of the House Foreign Affairs Committee March 14, 2019; approved by the House March 25, 2019.

The Prohibiting Unauthorized Military Action in Venezuela Act, H.R. 1004 (Cicilline), would prohibit funds made available to federal departments or agencies from being used to introduce the armed forces of the United States into hostilities with Venezuela, except pursuant to (1) a declaration of war, (2) a specific statutory authorization that meets the requirements of the War Powers Resolution and is enacted after the enactment of this bill, or (3) a national emergency created by attack upon the United States or the armed forces. Introduced February 6, 2019; referred to the House Foreign Affairs Committee and the House Armed Services Committee; reported out of the House Foreign Affairs Committee, as amended, April 9, 2019.

The Russian-Venezuelan Threat Mitigation Act, H.R. 1477 (Wasserman Schultz), would require the Secretary of State to submit an assessment within 120 days of its enactment on Russian-Venezuelan security cooperation and the potential threat such cooperation poses to the United States and countries in the Western Hemisphere. Introduced February 28, 2019; reported out of the House Foreign Affairs Committee March 14, 2019; approved by the House March 25, 2019.

The Venezuelan Contracting Restriction Act, H.R. 2204 (Waltz)/S. 1151(Scott), would prohibit an executive agency from entering into a contract for the procurement of goods or services with any person that has business relations with an authority of the Maduro government. Introduced in the House April 10, 2019; referred to the House Committee on Government Reform. Introduced in the Senate April 11, 2019; referred to the Senate Homeland Security and Government Affairs Committee.

The Department of State, Foreign Operations, and Related Programs Appropriations Act, FY2020, H.R. 2839 (Lowey), would provide $17.5 million in democracy and human rights aid to Venezuela. The report (H.Rept. 116-78) would direct Migration and Refugee Assistance and International Disaster Assistance to addressing the Venezuela migration crisis. Introduced and reported out of the Appropriations Committee May 20, 2019 (H.Rept. 116-78).

The Venezuela Temporary Protected Status Act of 2019, S. 636 (Menendez), would designate Venezuela under Section 244 of the

Immigration and Nationality Act to permit nationals of Venezuela to be eligible for Temporary Protected Status under such section. Introduced February 28, 2019; referred to the Judiciary Committee.

The VERDAD Act of 2019, S. 1025 (Menendez), would establish as U.S. policy the pursuit of a peaceful, diplomatic solution to the Venezuelan crisis; authorize $400 million in new humanitarian assistance; prohibit visas for the family members of sanctioned individuals but establish a waiver with conditions to lift visa restrictions; remove sanctions on designated individuals not involved in human rights abuses if they recognize Venezuela's interim president; require the State Department to work with Latin American and European governments to implement their own sanctions; require the Departments of State, Treasury, and Justice to lead international efforts to freeze, recover, and repurpose the corrupt financial holdings of Venezuelan officials; and accelerate planning with international financial institutions on the economic reconstruction of Venezuela contingent upon the restoration of democratic governance. Introduced April 22, 2019; reported with an amendment in the nature of a substitute to include three House measures (H.R. 854, H.R. 920, and H.R. 1477) out of the Senate Foreign Relations Committee May 22, 2019.

Resolutions

The Prohibiting Unauthorized Military Action in Venezuela Resolution of 2019, S.J.Res. 11 (Merkley), would prohibit U.S. department or agency funding from being used to introduce armed forces into hostilities with Venezuela, except pursuant to a specific statutory authorization by Congress enacted after this joint resolution. Introduced February 28, 2019.

Appendix B. Online Human Rights Reporting on Venezuela

Table B-1. Online Human Rights Reporting on Venezuela

Organization	**Document/Link**
Amnesty International	The State of the World's Human Rights, https://www.amnesty.org/en/countries/americas/venezuela/report-venezuela/
Committee to Protect Journalists	http://www.cpj.org/americas/venezuela/
Foro Penal Venezolano	http://foropenal.com/
Human Rights Watch	http://www.hrw.org/en/americas/venezuela
Inter-American Commission on Human Rights (IACHR)	http://www.cidh.oas.org/DefaultE.htm; Annual Report of the IACHR 2018, 2019, chapter IV includes a special report on Venezuela, http://www.oas.org/en/iachr/docs/annual/2018/TOC.asp
Programa Venezolano de Educación-Acción en Derechos Humanos (PROVEA)	http://www.derechos.org.ve/
Organization	Document/Link
Reporters Without Borders	https://rsf.org/en/venezuela
U.S. State Department	Country Report on Human Rights Practices 2018, March 13, 2019 https://www.state.gov/reports/2018-country reports-on-human-rights-practices/
Venezuelan Politics and Human Rights	Blog hosted by the Washington Office on Latin America, http://venezuelablog.tumblr.com

Source: Congressional Research Service.

In: Key Congressional Reports for June 2019 ISBN: 978-1-53616-666-8
Editor: Tara T. Robertson

Chapter 6

War Legacy Issues in Southeast Asia: Unexploded Ordnance (UXO)*

Michael F. Martin, Ben Dolven,
Andrew Feickert and Thomas Lum

Introduction

More than 40 years after the end of the Vietnam War, unexploded ordnance (UXO) from numerous conflicts, but primarily dropped by U.S. forces over Cambodia, Laos, and Vietnam during the Vietnam War, continues to cause casualties in those countries. Over the past 25 years, the United States has provided a total of over $400 million in assistance for UXO clearance and related activities in those three countries through the Department of Defense (DOD), Department of State (DOS), and United States Agency for International Development (USAID), as well as funding for treatment of victims through USAID and the Leahy War Victims fund. Although casualty numbers have dropped in recent years, no systematic

* This is an edited, reformatted and augmented version of Congressional Research Service, Publication No. R45749, dated June 3, 2019.

assessment of affected areas has been done, and many observers believe it may still take decades to clear the affected areas.

War legacy issues such as UXO clearance and victim assistance may raise important considerations for Congress as it addresses the impact of U.S. participation in conflicts around the world and how the United States should deal with the aftermath of such conflicts. The continued presence of UXO in Southeast Asia raises numerous issues, including appropriate levels of U.S. assistance for clearance activities and victim relief; coordination in efforts among DOD, DOS, and USAID; the implications of U.S. action on relations with affected countries; whether U.S. assistance in Southeast Asia carries lessons for similar activity in other parts of the world, including Iraq and Afghanistan; and, more generally, efforts to lessen the prevalence of UXO in future conflicts.

Many observers argue that U.S. efforts to address UXO issues in the region, along with joint efforts regarding other war legacy issues such as POW/MIA identification and Agent Orange/dioxin remediation, have been important steps in building relations with the affected countries in the post-war period. These efforts that have proceeded furthest in Vietnam, where the bilateral relationship has expanded across a wide range of economic and security initiatives. In Cambodia and Laos, where bilateral relations are less developed, UXO clearance is one of the few issues on which working-level officials from the United States and the affected countries have cooperated for years. Although some Cambodians and Laotians view U.S. demining assistance as a moral obligation and the U.S. government has viewed its support for UXO clearance as an important, positive aspect of its ties with the two countries, the issue of UXO has not been a major factor driving the relationships.

The Consolidated Appropriations Act, 2019 (P.L. 116-6) provides $196.5 million for "conventional weapons destruction" around the world, including $159.0 million for "humanitarian demining," with $3.85 million appropriated for Cambodia, $30.0 million for Laos, and $15.0 million for Vietnam.

The Legacies of War Recognition and Unexploded Ordnance Removal Act (H.R. 2097) would authorize $50 million per year for fiscal years 2020 to 2024 for humanitarian assistance in Cambodia, Laos, and Vietnam to develop national UXO surveys, conduct UXO clearance, and finance capacity building, risk education, and support for UXO victims.

OVERVIEW

The remnants of the Vietnam War (1963-1975) and other regional conflicts have left mainland Southeast Asia as a region heavily contaminated with unexploded ordnance, or UXO.[1] More than 45 years after the United States ceased its extensive bombing of Cambodia, Laos, and Vietnam, hundreds of civilians are still injured or killed each year by UXO from those bombing missions or by landmines laid in conflicts between Cambodia and Vietnam (1975-1978), China and Vietnam (1979-1990) and during the Cambodian civil war (1978-1991). While comprehensive surveys are incomplete, it is estimated that more than 20% of the land in Cambodia, Laos, and Vietnam are contaminated by UXO.

Over more than 25 years, Congress has appropriated more than $400 million to assist Cambodia, Laos, and Vietnam in clearing their land of UXO. More than 77% of the assistance has been provided via programs funded by the Department of State. In addition, the United States has provided treatment to those individuals maimed by UXO through U.S. Agency for International Development (USAID) programs and the Leahy War Victims Fund.

Despite ongoing efforts by the three countries, the United States, and other international donors, it reportedly could take 100 years or more, at the current pace, to clear Cambodia, Laos, and Vietnam of UXO.[2] During

---

[1] The Department of Defense (DOD) defines UXO as "Explosive ordnance that has been primed, fuzed, armed, or otherwise prepared for action, and which has been fired, dropped, launched, projected, or placed in such a manner as to constitute a hazard to operations, installations, personnel, or material and remains unexploded either by malfunction or design or for any other cause." See Joint Publication (JP) 3-15, "Barriers, Obstacles, and Mine Warfare for Joint Operations," March 5, 2018.

[2] See Padriac Convery, "200 Years to Go Before Laos Is Cleared of Unexploded US Bombs from Vietnam War Era," *South China Monring Post*, November 16, 2018; Vu Minh, "Decades

that time period, more people will likely be killed or injured by UXO. In addition, extensive areas of the three nations will continue to be unavailable for agriculture, industry, or habitation, hindering the economic development of those three nations.

In 2016, President Obama pledged $90 million over a three-year period for UXO decontamination programs in Laos—an amount nearly equal to the total of U.S. UXO assistance to that nation over the previous 20 years.[3] The Consolidated Appropriations Act, 2019 (P.L. 116 6) provides $196.5 million globally for "conventional weapons destruction," including $159.0 million for "humanitarian demining," under the Department of State's International Security Assistance programs.[4] Of the humanitarian demining funds, $3.85 million is appropriated for Cambodia, $30.0 million for Laos,[5] and $15.0 million for Vietnam. The act also provides $13.5 million for global health and rehabilitation programs under the Leahy War Victims Fund.

Moving forward, the 116th Congress will have an opportunity to consider what additional efforts, if any, the U.S. government should undertake to address the war legacy issue of UXO in mainland Southeast Asia in terms of the decontamination of the region and the provision of medical support or assistance to UXO victims. Beyond the immediate assistance such UXO-related programs would provide to Cambodia, Laos, and Vietnam, U.S. aid on this war legacy issue may also foster better bilateral ties to those nations. For example, some observers view U.S. assistance to Vietnam for the war legacy issue of Agent Orange/dioxin contamination as playing an important role in improving bilateral relations.[6]

After War, Vietnam Threatened by 800,000 Tons of Explosives," *VN Express*, April 2, 2018.

[3] Amy Sawitta Lefevre and Roberta Rampton, "U.S. Gives Laos Extra $90 Million to Help Clear Unexploded Ordnance," *Reuters*, September 6, 2016.

[4] The specific allocations are contained in H.Rept. 116-9, which accompanies the act.

[5] S.Rept. 115-282 to accompany S. 3108, the Department of State, Foreign Operations, and Related Programs Act, FY2020, and Joint Explanatory Statement to accompany P.L. 116-6, the Consolidated Appropriations Act, 2019.

[6] For more about U.S. Agent Orange/dioxin assistance to Vietnam, see CRS Report R44268, *U.S. Agent Orange/Dioxin Assistance to Vietnam*, by Michael F. Martin.

Background on Unexploded Ordnance

What Is Unexploded Ordnance (UXO)?

Unexploded ordnance (UXO) is defined as military ammunition or explosive ordnance which has failed to function as intended. UXO is also sometimes referred to as Explosive Remnants of War (ERW) or "duds" because of their failure to explode or function properly. UXO includes mines, artillery shells, mortar rounds, hand or rocket-propelled grenades, and rocket or missile warheads employed by ground forces (see Figure 1). Aerial delivered bombs, rockets, missiles, and scatterable mines that fail to function as intended are also classified as UXO. While many of these weapons employ unitary warheads, some weapons—primarily certain artillery shells, rocket and missile warheads and aerial bombs—employ cluster munitions, which disperse a number of smaller munitions as part of their explosive effect. Often times, these submunitions fail to function as intended. In addition, abandoned or lost munitions that have not detonated are also classified as UXO.

Source: George Black, "The Vietnam War Is Still Killing People," *The New Yorker*, May 20, 2016.

Figure 1. UXO Recovered in Vietnam, 2005.

The probability of UXO detonating is highly unpredictable; it depends on whether or not the munition has been fired, the level of corrosion or degradation, and the specific arming and fusing mechanisms of the device. "Similar items may respond very differently to the same action—one may be moved without effect, while another may detonate. Some items may be moved repeatedly before detonating and others may not detonate at all." In all cases, UXO poses a danger to both combatants and unaware and unprotected civilians.[7]

Military Use

Military munitions are used in a variety of ways. Some are used in direct force-on-force combat against troops, combat vehicles, and structures. Others, such as emplaced anti-personnel and anti-vehicle mines or scatterable mines, can be used to attack targets, deny enemy use of key terrain, or establish barriers to impede or influence enemy movement. Cluster munitions can either explode on contact once dispensed or can remain dormant on the ground until triggered by human or vehicular contact. The military utility of cluster weapons is that they can create large areas of destruction, meaning fewer weapons systems and munitions are needed to attack targets.

Mines and Cluster Munitions

Two particular classes of ordnance—mines and cluster munitions—have received a great deal of attention. Emplaced mines by their very nature pose a particular threat because they are often either buried or hidden and, unless their locations are recorded or some type of warning signs are posted, they can become easily forgotten or abandoned as the

---

[7] Landmine Action: The Campaign against Landmines, "Explosive Remnants of War: Unexploded Ordnance and Post-Conflict Communities," 2002, http://www.cpeo.org/pubs/UXOreport_3_26.pdf.

battlefield shifts over time. Cluster munitions are dispersed over an area and are generally smaller than unitary warheads, which can make them difficult to readily identify (see Figure 2). Since the conclusion of the Vietnam War, many of the newer mines and cluster munitions have a self-destruct or disarming capability. However, as long as their explosive charge remains viable, they pose a hazard to people.[8]

Both mines and cluster munitions have been subject to international protocols to limit or ban their development, transfer, and use. The 1999 Ottawa Convention "prohibits the use, stockpiling, production, and transfer of anti-personnel landmines (APLs). It requires states to destroy their stockpiled APLs within four years and eliminate all APL holdings, including mines currently planted in the soil, within 10 years."[9] The 2010 Convention on Cluster Munitions prohibits all use, stockpiling, production and transfer of cluster munitions. The United States has refused to sign either convention, citing the military necessity of these munitions.[10] The United States has, however, been a States Party to the Convention on the Use of Certain Conventional Weapons (CCW) since 1995, which "aims to protect military troops from inhumane injuries and prevent noncombatants from accidentally being wounded or killed by certain types of arms."[11] In 2009, the United States ratified Protocol V of the CCW, Explosive Remnants of War.[12] Protocol V "covers munitions, such as artillery shells, grenades, and gravity bombs, that fail to explode as intended, and any unused explosives left behind and uncontrolled by armed forces."[13] Under Protocol V "the government controlling an area with explosive remnants of

---

[8] For more about cluster munitions, see CRS Report RS22907, *Cluster Munitions: Background and Issues for Congress*, by Andrew Feickert and Paul K. Kerr.

[9] The Arms Control Association, The Ottawa Convention Fact Sheet," https://www.armscontrol.org/factsheets/ ottawasigs, updated January 2018.

[10] Cambodia is a signatory to, and Laos and Vietnam have signed but not ratified, the Ottawa Convention. Laos is a signatory to the Convention on Cluster Munitions, while Cambodia and Vietnam are not.

[11] Arms Control Association Fact Sheet, "Convention on Certain Conventional Weapons (CCW) at a Glance," https://www.armscontrol.org/factsheets/CCW, updated September 2017.

[12] International Committee on the Red Cross, "Treaties, States Parties, and Commentaries," https://ihldatabases.icrc.org/applic/ihl/ihl.nsf/States.xsp?xp_viewStates=XPages_NORMStatesParties&xp_treatySelected=610, accessed May 17, 2019.

[13] Arms Control Association Fact Sheet, "Convention on Certain Conventional Weapons (CCW) at a Glance," https://www.armscontrol.org/factsheets/CCW, updated September 2017.

war is responsible for clearing such munitions. However, that government may ask for technical or financial assistance from others, including any party responsible for putting the munitions in place originally, to complete the task. No state-party is obligated to render assistance."[14]

Source: Tom Fawthrop, "The Curse of Cluster Bombs," *Institute for Policy Studies*, September 30, 2011.

Figure 2. Unexploded Aerial Cluster Bomb and Submunitions.

The United States has undertaken a variety of initiatives—including mandating changes to munitions design and adopting federal safeguards and policy regulating their usage—to help limit the potential hazards posed to noncombatants by these UXO.[15]

U.S. Policy on Cluster Munitions

On June 19, 2008, then-Secretary of Defense Robert Gates issued a new policy on the use of cluster munitions.[16] The policy stated that "[c]luster munitions are legitimate weapons with clear military utility," but

---

[14] Ibid.

[15] For additional information on U.S. cluster munitions policy see CRS Report RS22907, *Cluster Munitions: Background and Issues for Congress*, by Andrew Feickert and Paul K. Kerr.

[16] Department of Defense, *DoD Policy on Cluster Munitions and Unintended Harm to Civilians*, June 19, 2008.

it also recognized "the need to minimize the unintended harm to civilians and civilian infrastructure associated with unexploded ordnance from cluster munitions." To that end, the policy mandated that after 2018, "the Military Departments and Combatant Commands will only employ cluster munitions containing submunitions that, after arming, do not result in more than 1% unexploded ordnance (UXO) across the range of intended operational environments."

On November 30, 2017, then-Deputy Secretary of Defense Patrick Shanahan issued a revised policy on cluster munitions.[17] The revised policy reverses the 2008 policy that established an unwaiverable requirement that cluster munitions used after 2018 must leave less than 1% of unexploded submunitions on the battlefield. Under the new policy, combatant commanders can use cluster munitions that do not meet the 1% or less unexploded submunitions standard in extreme situations to meet immediate warfighting demands. Furthermore, the new policy does not establish a deadline to replace cluster munitions exceeding the 1% rate, and these munitions are to be removed only after new munitions that meet the 1% or less unexploded submunitions standard are fielded in sufficient quantities to meet combatant commander requirements. However, the new DOD policy stipulates that the Department "will only procure cluster munitions containing submunitions or submunition warheads" meeting the 2008 UXO requirement or possessing "advanced features to minimize the risks posed by unexploded submunitions."

UXO IN SOUTHEAST ASIA

Overview

Although UXO in Southeast Asia can date back to World War II, the majority of the hazard is attributed to the Vietnam War. While an undetermined amount of UXO associated with the Vietnam War was from

[17] Deputy Secretary of Defense, "DoD Policy on Cluster Munitions," memorandum, November 30, 2017.

ground combat and emplaced mines, an appreciable portion of UXO is attributed to the air war waged by the United States from 1962 to 1973, considered by some to be one of the most intense in the history of warfare.[18] One study notes the United States dropped a million tons of bombs on North Vietnam. Three million more tons fell on Laos and Cambodia—supposedly "neutral" countries in the conflict. Four million tons fell on South Vietnam—America's ally in the war against communist aggression. When the last raid by B-52s over Cambodia on August 15, 1973, culminated American bombing in Southeast Asia, the United States had dropped more than 8 million tons of bombs in 9 years. Less than 2 years later, Cambodia, Laos, and South Vietnam were communist countries.[19]

The U.S. State Department in 2014 characterized the problem by country.[20]

- Cambodia: Nearly three decades of armed conflict left Cambodia severely contaminated with landmines and unexploded ordnance (UXO). The Khmer Rouge, the Royal Cambodian Armed Forces (RCAF), the Vietnamese military, and, to a lesser extent, the Thai army, laid extensive minefields during the Indochina wars. These minefields are concentrated in western Cambodia, especially in the dense "K-5 mine belt" along the border with Thailand, laid by Vietnamese forces during the 1980s. UXO—mostly from U.S. air and artillery strikes during the Vietnam War and land battles fought along the border with Vietnam—contaminates areas in eastern and northeastern Cambodia. While the full extent of contamination is unknown, the Landmine and Cluster Munition Monitor reports that a baseline survey completed in 2012 of

[18] Holy High, James R. Curran, and Gareth Robinson, "Electronic Records of the Air War over South East Asia: A Database Analysis," *Journal of Vietnamese Studies*, vol. 8, 2014, p. 89.

[19] Mark Clodfelter, "The Limits of Airpower or the Limits of Strategy: The Air Wars in Vietnam and Their Legacies," *Joint Forces Quarterly*, National Defense University, 3rd Quarter 2015, pp. 111-112.

[20] U.S. Department of State, *To Walk the Earth Safely: East Asia and Pacific*, 13th Edition, September 30, 2014.

Cambodia's 124 mine-affected districts found a total of 1,915 square kilometers (739 square miles) of contaminated land.

- Laos: Laos is the most heavily bombed country per capita in the world as a result of the Indochina wars of the 1960s and 1970s. While landmines were laid in Laos during this period, UXO, including cluster munitions remnants (called "bombies" in Laos), represents a far greater threat to the population and account for the bulk of contamination. UXO, mostly of U.S. origin, remains in the majority of the country's 18 provinces.
- Vietnam: UXO contaminates virtually all of Vietnam as a result of 30 years of conflict extending from World War II through the Vietnam War. The most heavily contaminated provinces are in the central region and along the former demilitarized zone (DMZ) that divided North Vietnam and South Vietnam. Parts of southern Vietnam and areas around the border with China also remain contaminated with UXO.

The Situation in Cambodia

The Kingdom of Cambodia is among the world's most UXO-afflicted countries, contaminated with cluster munitions, landmines, and other undetonated weapons. U.S. bombing of northeastern Cambodia during the Vietnam War, the Vietnamese invasion in 1979, and civil wars during 1970s and 1980s all contributed to the problem of unexploded ordnance. In 1969, the United States launched a four-year carpet-bombing campaign on Cambodia, dropping 2.7 million tons of ordnance, including 80,000 cluster bombs containing 26 million submunitions or bomblets.[21] Up to one-quarter of the cluster bomblets failed to explode, according to some

[21] The amount of U.S. ordnance dropped on Cambodia was more than the amount that the Allies dropped on Germany and Japan combined during World War II. See Henry Grabar, "What the U.S. Bombing of Cambodia Tells Us About Obama's Drone Campaign," *The Atlantic*, February 14, 2013; and Christopher Shay, "Cluster Munitions in the Spotlight as Groups Seek Ban on Bombs," *Phnom Penh Post*, October 30, 2008.

estimates.[22] In addition, the Vietnamese army mined the Cambodia-Thai border as it invaded the country and took control from the Khmer Rouge in 1979. The Vietnamese military, Vietnam-backed Cambodian forces, the Khmer Rouge, and Royalist forces reportedly all deployed landmines during the 1979-1989 civil war period.[23] Cambodian Prime Minister Hun Sen occasionally has referred to the U.S. bombing of Cambodia, which occurred between 1969 and 1973, when criticizing the United States;[24] however, the historical event has not been a major issue in recent U.S.-Cambodian relations.

Contamination and Casualties

There have been over 64,700 UXO casualties in Cambodia since 1979, including over 19,700 deaths.[25] The Cambodia Mine/ERW Victim Information System (CMVIS) has recorded an overall trend of significant decreases in the number of annual casualties: 58 in 2017 compared to 111 in 2015, 186 in 2012 and 286 in 2010.[26] Despite progress, the migration of poor Cambodians to the northwestern provinces bordering Thailand, one of the most heavily mined areas in the world, has contributed to continued casualties.[27] Cambodia, with 25,000 UXO-related amputees, has the highest number of amputees per capita in the world.[28] The economic costs of UXO include obstacles to infrastructure development, land unsuitable for agricultural purposes, and disruptions to irrigation and drinking water supplies.

---

[22] Christopher Shay, "Cluster Munitions in the Spotlight as Groups Seek Ban on Bombs," *Phnom Penh Post*, October 30, 2008; Landmine and Cluster Munition Monitor, *Cambodia, Mine Action*, http://www.the-monitor.org/en-gb/reports/ 2018/cambodia/mine-action.aspx.

[23] The Halo Trust, *Cambodia*, https://www.halotrust.org/where-we-work/south-asia/cambodia/.

[24] Kimkong Heng, "Is Cambodia's Foreign Policy Heading in the Right Direction?" *The Diplomat*, February 8, 2019.

[25] Landmine and Cluster Munition Monitor, *Cambodia, Casualties*, http://the-monitor.org/en-gb/reports/2018/ cambodia/casualties.aspx.

[26] Ibid.

[27] Department of State, *To Walk the Earth in Safety*, 16th Edition, December 13, 2017; Peter Zsombor, "Cambodia Sees Its First Month Without Mine, UXO Victims," *The Cambodia Daily*, July 28, 2017.

[28] The Halo Trust, op. cit.

Open Development Cambodia, a website devoted to development-related data, reports that since the early 1990s, about 580 square miles (1,500 square kilometers) of land has been cleared of UXO.[29] Estimates of the amount of land still containing UXO vary. According to some reports, about 50% of contaminated land has been cleared, and an estimated 630 square miles (1,640 square kilometers) of land still contain UXO.[30] Many of the remaining areas are the most densely contaminated, including 21 northwestern districts along the border with Thailand that contain anti-personnel mines laid by the Vietnamese military and that account for the majority of mine casualties.[31]

Cleanup Efforts

Between 1993 and 2017, the U.S. government contributed over $133.6 million for UXO removal and disposal, related educational efforts, and survivor assistance programs in Cambodia. These activities are carried out largely by U.S. and international nongovernmental organizations (NGOs), in collaboration with the Cambodian Mine Action Center, a Cambodian NGO, and the Cambodian government. USAID's Leahy War Victims Fund has supported programs to help provide medical and rehabilitation services and prosthetics to Cambodian victims of UXO.[32] Nonproliferation, Anti-terrorism, Demining and Related Programs (NADR) funding for demining activities was $5.5 million in both 2015 and 2016, $4.2 in 2017, and $2.9 million in 2018.[33]

---

[29] Open Development Cambodia, Landmines and Unexploded Ordnance in Cambodia, https://opendevelopmentcambodia.net/topics/landmines-uxo-and-demining/#return-note-73803-21.

[30] Landmine and Cluster Munition Monitor, *Cambodia, Mine Action*, op. cit.; Michael Hart, "Can Cambodia Meet Its Target to Remove Landmines by 2025?" *Asian Correspondent*, December 4, 2017.

[31] Ibid.

[32] Department of State, *To Walk the Earth in Safety*, 17th Edition, December 18, 2018.

[33] Nonproliferation, Anti-terrorism, Demining, and Related Programs-Conventional Weapons Destruction (NADRCWD). Data from Department of State.

Global donors contributed over $132 million between 2013 and 2017, mostly for clearance efforts. In 2017, the largest contributors of demining and related assistance were the United States, United Kingdom, Australia, Japan, and Germany, providing approximately $10.6 million in total.[34]

In 2018, the Cambodian government and Cambodian Mine Action and Victim Assistance Authority (CMAA), a government agency, launched the National Mine Action Strategy (NMAS) for 2018-2025. The goal of removing UXO from all contaminated areas by 2025 would require the clearance of 110 square kilometers per year at a cost of about $400 million.[35] The NMAS estimated that at the current rate of progress, however, Cambodia would need a little over 10 years to complete clearance of all known mined areas. Some experts are concerned that declining international assistance could jeopardize clearance goals.[36] In 2017, total international demining support to Cambodia decreased by 61%, largely due to lower contributions from Australia and Japan.[37]

The Situation in Laos

From 1964 through 1973, the United States military reportedly flew 580,000 bombing runs and dropped over 2 million tons of cluster munitions, including over 270 million cluster bombs, on the small land-locked country.[38] The total was more than the amount dropped on Germany and Japan combined in World War II. An estimated one-third of these munitions failed to explode.[39] The Lao government claims that up to 75-80 million submunitions or bomblets released from the cluster bombs

---

[34] Landmine and Cluster Munition Monitor, *Cambodia, Support for Mine Action*, http://www.the-monitor.org/en-gb/ reports/2018/cambodia/support-for-mine-action.aspx.

[35] Landmine and Cluster Munition Monitor, *Cambodia, Support for Mine Action*, ibid.; Pech Sotheary, "China Grants Millions for Mine Clearance," *Khmer Times*, May 28, 2018.

[36] Department of State, *To Walk the Earth in Safety*, 17th Edition, December 18, 2018.

[37] Landmine and Cluster Munition Monitor, *Cambodia, Support for Mine Action*, op. cit.

[38] National Regulatory Authority for UXO/Mine Action Sector for Lao PDR, *Annual Report 2015*, and Legacies of War, "Secret War in Laos," http://legaciesofwar.org/about-laos/secret-war-laos/.

[39] White House, "Remarks by President Obama to the People of Laos," September 6, 2016.

remain in over one-third of the country's area.[40] Military conflicts during the French colonial period and the Laotian Civil War during the 1960s and 1970s have also contributed to the problem of UXO/ERW.

The U.S. bombing campaign in Laos was designed to interdict North Vietnamese supply lines that ran through Laos. The bombing campaign also supported Lao government forces fighting against communist rebels (Pathet Lao) and their North Vietnamese allies. Cluster munitions were considered the "weapon of choice" in Laos because they could penetrate the jungle canopy, cover large areas, and successfully attack convoys and troop concentrations hidden by the trees. The most heavily bombed areas in Laos were the northeastern and southern provinces, although UXO can be found in 14 of the country's 17 provinces. The bombings in the northeast were intended to deny territory, particularly the Plain of Jars, to Pathet Lao and North Vietnamese forces and, in the south, to sever the Ho Chi Minh Trail, which crossed the border into eastern Laos. The northeastern part of Laos was also used as a "free drop zone" where planes that had taken off from bases in Thailand and had been unable to deliver their bombs, could dispose of them before returning to Thailand.

Contamination and Casualties

According to the Geneva-based *Landmine and Cluster Munition Monitor*, since 1964, there have been over 50,000 mine and ERW casualties in Laos, including over 29,000 people killed.[41] An estimated 40% of victims are children.[42] In 2012, the Lao government's *Safe Path Forward Strategic Plan II* set a target to reduce UXO-related casualties to 75 per year by 2020, from levels between 100-200 victims annually during

[40] Legacies of War, "Leftover Unexploded Ordnance (UXO)," http://legaciesofwar.org/about-laos/leftover-unexplodedordnances-uxo/.

[41] Landmine and Cluster Munitions Monitor, "Lao PDR: Casualties and Victim Assistance," updated June 5, 2015.

[42] CNN, "'My Friends Were Afraid of Me': What 80 Million Unexploded US Bombs Did to Laos," September 6, 2016.

the 2000s. The country has already met these goals: in 2017, the number of reported casualties was 41, including four killed.

Cluster munitions have hampered economic development in the agricultural country. UXO contamination affects one-quarter of all Lao villages, and 22% of detonations occur through farming activities.[43] Unexploded ordnance adversely affects not only agricultural production, but also mining, forestry, the development of hydropower projects, and the building of roads, schools, and clinics. Expenditures on demining efforts and medical treatment divert investment and resources from other areas and uses. Many injured UXO survivors lose the ability to be fully productive. According to the Lao government, there appears to be a significant correlation between the presence of UXO and the prevalence of poverty.[44]

Cleanup Efforts

Lao PDR officials state that the country needs $50 million annually for ongoing UXO/ERW clearance, assistance to victims, and education, of which the Lao government contributes $15 million.[45] International assistance comes from numerous sources, including Japan, the United States, and the United Nations Development Program (UNDP). The United States has contributed a total of $169 million for UXO clearance and related activities since 1995, with funding directed to international NGOs and contractors.[46] That makes Laos the third largest recipient of conventional weapons destruction funding over that period, after Afghanistan and Iraq.[47] In 2016, the United States announced a three-year, $90 million increase in assistance covering FY2016- FY2018. Half the amount, or $45 million, is aimed at conducting the first nationwide cluster

---

[43] Ibid.

[44] National Regulatory Authority for UXO/Mine Action Sector for Lao PDR, op. cit.

[45] Khonesavanh Latsaphao, "Laos Is Not Meeting UXO Clearance Targets," *Vientiane Times*, May 3, 2013.

[46] Department of State, "Special Report: U.S. Conventional Weapons Destruction in Laos," July 17, 2018.

[47] Department of State, *To Walk the Earth in Safety,* 17th Edition, December 18, 2018.

munitions remnant survey, while the other half is aimed at clearance activities.

Since the early 1990s, the U.S. Department of Defense (DOD) has been involved in training Lao personnel in demining techniques. U.S. UXO clearance and related humanitarian aid efforts, administered by the State Department (DOS), began in 1996. U.S. support also helped to establish the Lao National Demining Office, the UXO Lao National Training Center, and the Lao National Regulatory Authority.

The United States finances the bulk of its mine clearance operations through the NADR foreign aid account. NADR demining programs constitute the largest U.S. assistance activity in Laos, which receives little U.S. development aid compared to other countries in the region. It has been channeled primarily to international nongovernmental organizations (NGOs), the UNDP's trust fund for UXO clearance, and the Lao National Unexploded Ordnance Program (UXO Lao). Laos also has received humanitarian assistance through the USAID Leahy War Victims Fund for prosthetics, orthotics, and rehabilitation ($1.4 million in 2011-2013).

For many years in the 1990s and 2000s, UXO-related clearance programs were one of the primary areas of substantive cooperation between the United States and Laos. Some argue that such activity has helped foster bilateral ties with a country whose authoritarian government is deeply inward looking. When President Obama became the first U.S. President to visit Laos in 2016, announcing the $90 million UXO aid package, he said: "Given our history here, I believe that the United States has a moral obligation to help Laos heal. And even as we continue to deal with the past, our new partnership is focused on the future."[48]

The Situation in Vietnam

War Legacy issues—Agent Orange/dioxin contamination, MIAs, and UXO—played an important role in the reestablishment of diplomatic relations between the United States and Vietnam, and it led to the

[48] White House, "Remarks by President Obama to the People of Laos," September 6, 2016.

development of a comprehensive partnership between the two nations. Vietnam's voluntary effort to locate and return the remains of U.S. MIAs was a significant factor in the restoration of diplomatic relations.[49] U.S. assistance to decontaminate Da Nang airport of Agent Orange/dioxin likely contributed to the two nations' move to a comprehensive partnership.[50] While not as prominent, U.S. UXO assistance to Vietnam most likely has been a factor in establishing trust between the two governments.

The UXO in Vietnam are remnants from conflicts spanning more than a century, potentially as far back as the Sino-French War (or Tonkin War) of 1884-1885 and as recent as the Cambodian-Vietnamese War (1975-1978) and the border conflicts between China and Vietnam from 1979 to 1991.[51] According to one account, during Vietnam's conflicts with France and the United States (1945-1975), more than 15 million tons of explosives were deployed—four times the amount used in World War II.[52] It is generally presumed, however, that the majority of the UXO in Vietnam are from the Vietnam War, also known in Vietnam as "the Resistance War Against America" (1955-1975).

Contamination and Casualties

Estimates of the amount of UXO in Vietnam vary. According to one source, "at least 350,000 tons of live bombs and mines remain in

---

[49] See CRS Report RL33316, *U.S.-Vietnam Relations in 2008: Background and Issues for Congress*, by Mark E. Manyin.

[50] For more about U.S. Agent Orange/dioxin assistance to Vietnam, see CRS Report R44268, *U.S. Agent Orange/Dioxin Assistance to Vietnam*, by Michael F. Martin.

[51] The Sino-French War occurred between August 1884 and April 1885 over control of what is now northern Vietnam, but known as Tonkin at the time. Although the combined Vietnamese and Chinese forces were ultimately victorious in the land conflict in Vietnam, France was able to secure control over Annam (central Vietnam) and Tonkin in the Treaty of Tientsin. In the Cambodian-Vietnamese war, although the Communist Party of Kampuchea (also known as the "Khmer Rouge") and the Socialist Republic of Vietnam (i.e., North Vietnam) were allies during the Vietnam War, fighting broke out between the newly-formed Democratic Kampuchea and Vietnam in 1975, ending with Vietnam's invasion of Cambodia in December 1978. China and Vietnam engaged in a series of border conflicts between 1979 and 1991, largely over disputed territory.

[52] Vu Minh, "Decades After War, Vietnam Threatened by 800,000 Tons of Explosives," *VN Express*, April 2, 2018.

Vietnam."[53] Another source claims "around 800,000 tons of unexploded ordnance remains scattered across the country."[54]

Viewed in terms of land area, the Vietnamese government estimates that between 6.1 and 6.6 million hectares (23,500-25,500 square miles) of land in Vietnam—or 19% to 21% of the nation—is contaminated by UXO.[55] An official Vietnamese survey started in 2004 and completed in 2014 estimated that 61,308 square kilometers (23,671 square miles) was contaminated with UXO.[56] According to the survey, UXO is scattered across virtually all of the nation, but the province of Quang Tri, along the previous "demilitarized zone" (DMZ) between North and South Vietnam, is the most heavily contaminated (see Figure 3).

Figures on the number of UXO casualties in Vietnam also vary. One source says, "No one really knows how many people have been injured or killed by UXO since the war ended, but the best estimates are at least 105,000, including 40,000 deaths."[57] In its report on UXO casualties in Vietnam, however, the Landmine and Cluster Munition Monitor listed the casualty figures for 1975-2017 as 38,978 killed and 66,093 injured.[58] For 2017 only, the Landmine and Cluster Munition Monitor reported eight deaths and six injured.

A survey of UXO casualties determined that the three main circumstances under which people were killed or injured by UXO were (in order): scrap metal collection (31.2%); playing/tampering (27.6%); and cultivating or herding (20.3%). In some of Vietnam's poorer provinces,

---

[53] Ariel Garfinkel, "The Vietnam War Is Over. The Bombs Remain," *New York Times*, March 20, 2018.

[54] Vu Minh, "Decades After War, Vietnam Threatened by 800,000 Tons of Explosives," *VN Express*, April 2, 2018.

[55] Le Ha, "Vietnam to Issue Unexploded Ordnance Map," *VietNamNet*, June 16, 2016; and Vu Minh, "Decades After War, Vietnam Threatened by 800,000 Tons of Explosives," *VN Express*, April 2, 2018.

[56] Vietnam National Mine Action Center, *Report on Explosive Remnants of War Contamination in Vietnam*, April 3, 2018.

[57] Vuong Duc Anh and Xavier Bourgois, "Vietnam's Never-Ending War: Into the Trenches with the Bomb Disposal Squad," *VN Express*, October 17, 2016.

[58] Landmine and Cluster Munitions Monitor, *Vietnam: Casualties*, October 10, 2018.

people proactively seek out and collect UXO in order to obtain scrap metal to sell to augment their income, despite the inherent danger.[59]

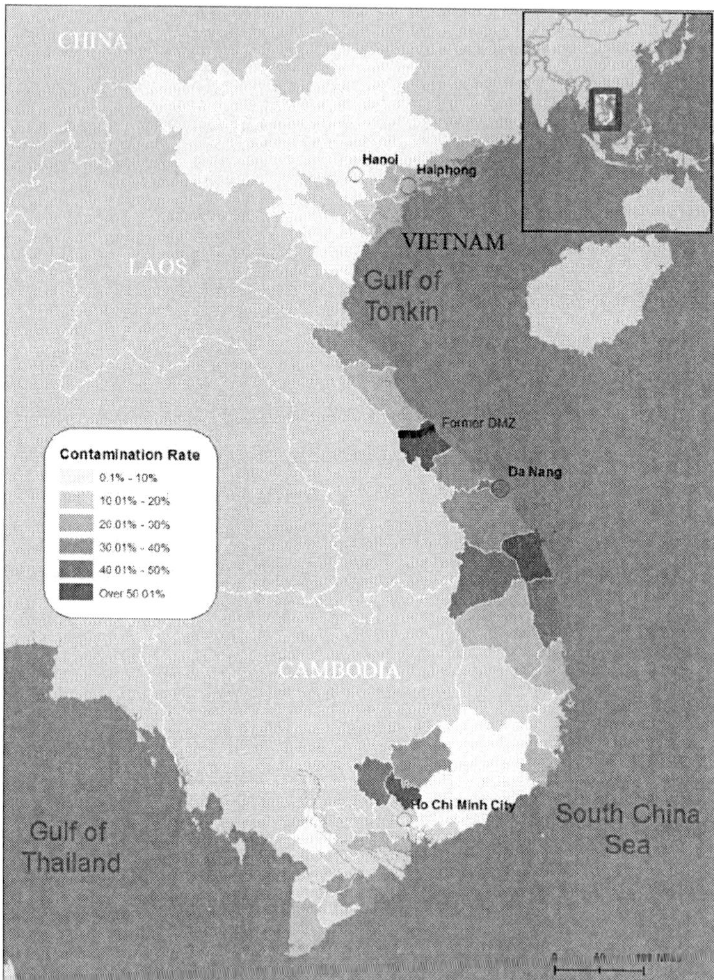

Source: CRS.

Notes: Based on data from *Report on Explosive Remnants of War Contamination in Vietnam*.

Figure 3. Unexploded Ordnance (UXO) Contamination in Vietnam. By province or city; percentage of land area contaminated.

[59] Hoang Tao, "Scrap Collectors' Dance with Death in Central Vietnam's Former Battlefields," *VN Express*, December 11, 2018.

Cleanup Efforts

On March 8, 2018, Vietnam's Ministry of National Defence (MND) established the Office of the Standing Agency of the National Steering Committee of the Settlement of Post-war Unexploded Ordnance and Toxic Chemical Consequences, or Office 701, to address the nation's UXO issue.[60] Office 701 is responsible for working with individuals and organizations to decontaminate Vietnam of UXO to ensure public safety, clean the environment, and promote socio-economic development. Under a 2013 directive by the Prime Minister, the Vietnam National Mine Action Center (VNMAC) was established within the MND with responsibility for proposing policy, developing plans, and coordinating international cooperation for UXO clearance.[61] The MND's Center for Bomb and Mine Disposal Technology (BOMICEN) is the central coordinating body for Vietnam's UXO clearance operations. In addition, Vietnam created a Mine Action Partnership Group (MAPG) to improve coordination of domestic and international UXO clearance operations.

BOMICEN typically sets up project management teams (PMTs) that work with provincial or local officials to identify, survey, and decontaminate UXO.[62] The PMTs usually interview local informants about possible UXO sites and then conduct field evaluations to determine if UXO is present and suitable for removal by Vietnam's Army Engineering Corps. The PMTs also collect information about the decontamination site and report back to BOMICON about the location and type of UXO removed.

Besides the clearance operations directly conducted by Vietnam, several nations and international organizations conduct UXO removal projects in Vietnam, including the Danish Demining Group (DDG), the Mines Advisory Group (MAG), Norwegian People's Aid (NPA), and PeaceTrees Vietnam. In 2016, the Korea International Cooperation Agency

---

[60] "New Office on Post-War Bomb, Mine Recovery Set Up," *VN Express*, March 9, 2018.

[61] Mine Action, *Vietnam*, November 16, 2018.

[62] Vietnam National Mine Action Center, *Report on Explosive Remnants of War Contamination in Vietnam*, 2018.

(KOICA), in cooperation with VNMAC and the United Nations Development Programme (UNDP), initiated a $32 million, multi-year UXO project in the provinces of Binh Dinh and Quang Binh.[63] The joint project began operations in March 2018 and is scheduled to end in December 2020.[64]

Table 1. U.S. UXO Assistance to Southeast Asia by country and source, in thousands of U.S. dollars

	2015	2016	2017	FY1993-FY2017[a]	2018	2019 est.	2020 req.
Cambodia							
* CDC	0	0	0	100	—	—	—
* DOD	2,379	1,717	1,969	24,063	—	—	—
* DOS	8,307	8,522	4,300	94,388	2,000	3,850	7,000
* USAID	500	303	0	15,084	—		
Subtotal				133,635			
Laos							
* CDC	0	0	0	0	—	—	—
* DOD		111	10	7,021	—		
* DOS	26,880	20500	30000	145,114	30,000	30,000	10,000
* USAID	2,000	2,166	3,005	16,971	—	—	—
Subtotal				169,106			
Vietnam							
* CDC	0	0	0	1,848	—	—	—
* DOD	340	722	1,168	4,295	—	—	—
* DOS	12,548	10,709	12,500	86,359	12,500	15,000	8,000
* USAID	0	0	0	26,799	—	—	—
Subtotal				119,301			
Total			422,042				

Source: FY2015-FY2017 numbers and FY1993-FY2017 totals: U.S. Department of State, *To Walk the Earth in Safety*, 17th Edition, December 18, 2018. FY2018 and FY2020 (requested) numbers: Department of State, *Congressional Budget Justification for Foreign Operations, FY2020*. FY2019 numbers (estimated): S.Rept. 115-282 to accompany S. 3108, the Department of State, Foreign Operations, and Related Programs Act, FY2020 and Joint Explanatory Statement to accompany P.L. 116-6, the Consolidated Appropriations Act, 2019.

Notes: [a]. Cambodia totals are from FY1995 to FY2017.

NGOs working in Vietnam report some issues in their collaboration with the MND, which has declared portions of contaminated provinces off limits for UXO surveying and decontamination. Many of these areas

[63] "Korea Offer $20 Mln to Help Vietnam Remove Unexploded Ordnance," *VN Express*, June 14, 2016.

[64] "Vietnam-S Korea Mine Cleanup Launched," *Viet Nam News*, March 10, 2018.

contain villages and towns inhabited by civilians. In addition, the MND has not been providing information about any UXO clearance efforts being conducted in these areas. The lack of information sharing has hindered efforts to establish a nationwide UXO database that is being used to refine UXO location and clearance techniques.[65]

U.S. UXO Assistance in Southeast Asia

Since 1993, the United States has provided UXO and related assistance to Southeast Asia via several different channels, including the Center for Disease Control (CDC), the Department of Defense (DOD), the Department of State (DOS), and the U.S. Agency for International Development (USAID) (see Table 1). For all three countries covered by this chapter, most of the assistance has been provided by DOS through its Nonproliferation, Anti-terrorism, Demining and Related Programs/ Conventional Weapons Destruction (NADR-CWD) account. USAID assistance to Cambodia, Laos, and Vietnam has consisted primarily of Leahy War Victims Fund programs for prostheses, physical rehabilitation, training, and employment. Laos, Cambodia, and Vietnam have been the largest recipients of U.S. conventional weapons destruction (CWD) funding in East Asia.

In December 2013, the United States and Vietnam signed a Memorandum of Understanding on cooperation to overcome the effects of "wartime bomb, mine, and unexploded ordnance" in Vietnam.[66] In their November 2017 joint statement, President Trump and President Tran Dai Quang "committed to cooperation in the removal of remnants of explosives from the war."[67]

---

[65] Based on CRS research trip to Vietnam, April 12-23, 2019.

[66] "Vietnam, US Sign First MOU on Bomb, Mine Clearance," *Vietnam News Brief Service*, December 17, 2013.

[67] White House, "Joint Statement: Between the United States of America and the Socialist Republic of Viet Nam," press release, November 13, 2017.

U.S. Department of State and USAID Activities

Department of State and USAID demining and related assistance support the work of international NGOs in Cambodia, Laos, and Vietnam. International NGOs work primarily with local NGOs in Cambodia and, to a greater extent, collaborate with government entities in Laos and Vietnam. The main areas of assistance are clearance, surveys, and medical assistance. In Cambodia, the Department of State and USAID support programs that collaborate with and train Cambodian organizations in clearance activities, conduct geographical surveys, help process explosive material retrieved from ERW, and provide mine risk education. In Laos, U.S. assistance includes clearance and survey efforts, medical and rehabilitation services, education and training assistance to victims and families, and mine risk education. In Vietnam, the United States provides mine clearance and survey support, capacity building programs, and medical assistance and vocational training for victims.[68]

U.S. Department of Defense (DOD) and UXO Remediation Activities

DOD's role in remediating UXO in Southeast Asia falls under the category of "Support to Humanitarian Mine Action (HMA)." Chairman of the Joint Chiefs of Staff (CJCS) Instruction "Department of Defense Support to Humanitarian Mine Action, CJCSI 3207.0IC" dated September 28, 2018, covers DOD's responsibilities in this regard. DOD's stated policy is

> to relieve human suffering and the adverse effects of land mines and other explosive remnants of war (ERW)[69] on noncombatants while

[68] Department of State, *To Walk the Earth in Safety*, 17th Edition, December 18, 2018; Department of State, *Congressional Budget Justifications for Foreign Operations* (multiple years); Landmine and Cluster Munition Monitor; USAID, *Foreign Aid Explorer*.

[69] DOD defines Explosive Remnants of War (ERW) as land mines, unexploded ordnance (UXO) (mortar rounds, artillery shells, bomblets, rockets, submunitions, rocket motors and fuel,

> advancing the Combatant Commanders' (CCDRs') *theater campaign plan* and U.S. national security objectives. The DOD HMA program assists nations plagued by land mines and ERW by executing "train-the-trainer" programs of instruction designed to develop indigenous capabilities for a wide range of HMA activities.[70]

It is important to note that U.S. Code restricts the extent to which U.S. military personnel and DOD civilian employees can actively participate in UXO activities as described in the following section:

Exposure of USG Personnel to Explosive Hazards

> By law, DOD personnel are restricted in the extent to which they may actively participate in ERW clearance and physical security and stockpile management (PSSM) operations during humanitarian and civic assistance. Under 10 U.S.C. 401(a)(1), Military
>
> Departments may carry out certain "humanitarian and civic assistance activities" in conjunction with authorized military operations of the armed forces in a foreign nation. 10 U.S.C. 407(e)(1) defines the term "humanitarian demining assistance" (as part of humanitarian and civic assistance activities) as "detection and clearance of land mines and other ERW, and includes the activities related to the furnishing of education, training, and technical assistance with respect to explosive safety, the detection and clearance of land mines and other ERW, and the disposal, demilitarization, physical security, and stockpile management of potentially dangerous stockpiles of explosive ordnance." However, under 10 U.S.C. 407(a)(3), members of the U.S. Armed Forces while providing humanitarian demining assistance shall not "engage in the physical detection, lifting, or destroying of land mines or other explosive remnants of war, or stockpiled conventional munitions (unless the member does so for the concurrent purpose of supporting a United States military operation)." Additionally, members of the U.S. Armed Forces shall not provide such humanitarian demining and civic assistance "as part of a

---

grenades, small-arms ammunition, etc.), and abandoned ammunition storage and cache sites.

[70] Chairman of the Joint Chiefs of Staff Instruction "Department of Defense Support to Humanitarian Mine Action, CJCSI 3207.01C," dated September 28, 2018, p. 1.

military operation that does not involve the armed forces." Under DOD policy, the restrictions in 10 U.S.C. 407 also apply to DOD civilian personnel.[71]

In general terms, U.S. law restricts DOD to "train-the-trainer" type UXO remediation activities unless it is required as part of a U.S. military operation involving U.S. armed forces.

U.S. Indo Pacific Command and UXO Remediation in Vietnam, Cambodia, and Laos[72]

U.S. Indo Pacific Command (USINDOPACOM)[73] is responsible for U.S. military activities in Vietnam, Cambodia, and Laos. As part of USINDOPACOM's Theater Campaign Plan,[74] selected UXO remediation activities for Vietnam, Cambodia, and Laos are briefly described in the following sections:

Vietnam

USINDOPACOM has tasked U.S. Army Pacific (USARPAC) as the primary component responsible for land-based UXO operations and the Pacific Fleet (PACFLT) as the primary component responsible for underwater UXO operations in Vietnam. FY2018 accomplishments and FY2019 and FY2020 plans are said to include:

---

[71] Ibid., p. 2.

[72] Information in this section was provided to CRS by USINDOPACOM on January 28, 2019.

[73] U.S. Pacific Command (USPASCOM) was renamed U.S. Indo Pacific Command (USINDOPACOM) in 2018 to reflect the growing role of India in U.S. national security.

[74] The Defense Security Cooperation Agency (DSCA) defines a theater campaign plan as: "Plans developed by geographic combatant commands that focus on the command's steady-state activities, which include operations, security cooperation, and other activities designed to achieve theater strategic end states."

FY2018

Trained individuals on International Mine Action Standards (IMAS)[75] Level I and II; Trained individuals on Explosive Ordnance Disposal (EOD) instructor development; Familiarized individuals on EOD equipment;

Conducted medical first responder training;

Trained individuals on medical instructor development;

Trained individuals on underwater remote vehicle operations; and

Trained individuals on ordnance identification.

FY2019

Continue training on International Mine Action Standards Level I;

Train individuals on how to develop training lanes for demining;

Exercise IMAS Level I concepts;

Increase Vietnamese medical first responder force structure; and

Continue EOD instructor development.

FY2020

Plan to train on IMAS Level II with qualified IMAS Level I students;

Plan to enhance advanced medical-related technique training;

Plan to train in demolition procedures;

Plan to train in maritime UXO techniques;

Plan to conduct mission planning and to conduct a full training exercise; and

Plan to conduct instructor development.

---

[75] According to the Geneva International Centre for Humanitarian Demining, International Mine Action Standards (IMAS) provide guidance, establish principles and, in some cases, define international requirements and specifications. They are designed to improve safety, efficiency and quality in mine action, and to promote a common and consistent approach to the conduct of mine action operations. The IMAS are intended to be the main guide for the development of National Mine Action Standards (NMAS), Standard Operating Procedures (SOPs), and training material in mine action. https://www.gichd.org/topics/standards/international-standards/#.XFBciVxKiUk, accessed January 29, 2019.

Cambodia

USINDOPACOM has tasked Marine Forces Pacific (MARFORPAC) to be responsible for land-based UXO operations in Cambodia. Plans for FY2019 through FY2021 include:

FY2019

Train in IMAS EOD Level I;

Train on EOD instructor development;

Familiarize students on EOD Level I equipment;

Review medical first responder training;

Train on medical instructor development;

Train in ordnance identification; and

Train in IMAS Demining Non-Technical Survey/Technical Survey (NTS/TS) techniques.

FY2020

Plan to continue to develop capacity with IMAS EOD Level I and II training;

Plan to continue to build capacity with IMAS Demining Non-Technical Survey/Technical Survey techniques;

If EOD Level I and II training successful, plan to initiate EOD Level III training in late FY2020;

Plan to increase student knowledge of lane training development;

Plan to exercise IMAS Level II concepts;

Plan to increase Cambodian medical first responder force structure; and

Plan to continue EOD instructor development.

FY2021

Plan to train on IMAS EOD Level II and EOD Level III with the qualified Level I and Level II students to increase their numbers;

Plan to train on IMAS Demining NTS/TS with the qualified students to increase their numbers;

Plan to enhance advanced medical-related techniques;

Plan to train in demolition procedures;

Plan to conduct mission planning and a full training exercise; and Plan to conduct instructor development events.

Laos

USINDOPACOM has tasked Marine Forces Pacific (MARFORPAC) to be responsible for land-based UXO operations in Laos. Plans for FY2019 through FY2021 include:

FY2019

Conduct training on IMAS EOD Level I;

Conduct training on EOD instructor development;

Conduct familiarization on EOD Level I equipment;

Conduct a review of medical first responder training;

Conduct medical instructor development training; and

Conduct training on ordnance identification.

FY2020

Plan to continue to build capacity with training in IMAS EOD Level I and II;

Plan to increase knowledge on lane training development;

Plan to exercise IMAS Level II concepts;

Plan to increase medical first responder force structure and knowledge; and

Plan to continue EOD instructor development.

FY2021

Plan to train on IMAS EOD Level II with the qualified Level I and Level II students to increase their numbers;

Plan to enhance advance medical-related techniques; Plan to train in demolition procedures;

Plan to conduct mission planning and conduct a full training exercise; and

Continue to conduct instructor development.

IMPLICATIONS FOR CONGRESS

The U.S. government has been providing UXO-related assistance to Southeast Asia for over 25 years, with contributions amounting to over $400 million. Despite this sustained level of support, as well as the efforts of the governments of Cambodia, Laos, and Vietnam, it may take decades to clear these three nations of the known UXO contamination. These estimates, however, are based on incomplete information, as systematic nationwide UXO surveys have not been completed in either Cambodia or Laos.

The Legacies of War Recognition and Unexploded Ordnance Removal Act (H.R. 2097) would authorize $50 million each year for fiscal years 2020 to 2024 for address the UXO issue in Cambodia, Laos, and Vietnam. The legislation also would authorize the President to provide humanitarian assistance for developing national UXO surveys, UXO clearance, and support for capacity building, risk education and UXO victims assistance in each nation. It would require the President to provide an annual briefing on related activities to the House Committee on Appropriations, the House Committee on Foreign Affairs, the Senate Committee on Appropriations, and the Senate Committee on Foreign Relations.

Southeast Asia's ongoing UXO challenge may present a number of issues for Congress to consider and evaluate. Among those issues are

- Funding levels—It is uncertain how much money it would take to decontaminate all three nations or provide adequate assistance to their UXO victims. Given this uncertainty, is the level of U.S. assistance being provided to Cambodia, Laos, and Vietnam to conduct humanitarian demining projects adequate to significantly reduce the UXO casualty risk in a reasonable time period? In addition, is the recent distribution of funding across the three nations equitable given their relative degrees of UXO contamination and their internal ability to finance demining projects?

- Coordination across agencies—Is there appropriate coordination across the U.S. agencies—the Department of Defense, the Department of State, and USAID—in providing demining assistance in Southeast Asia? Are these agencies utilizing the appropriated funds efficiently and effectively?
- Focus on clearance—Most of the appropriated funds have been for humanitarian demining projects and technical support, with less funding for assistance to UXO victims. The focus on clearance, rather than assistance on UXO victims, may in part be due to a concern about possible post-conflict liability issues. In light of past practices, should the U.S. government increase its support for UXO victims in Cambodia, Laos, and Vietnam beyond those being currently provided via the Leahy War Victims Fund?
- Implications for bilateral relations—Has the amount and types of U.S. UXO assistance to Cambodia, Laos, and Vietnam been a significant factor in bilateral relations with each of those nations? In Vietnam, work on war legacy issues formed an early part of building normalized relations in the post-War period— ties that have broadened into closer strategic and economic linkages. In Cambodia and Laos, UXO-related assistance has been one of the broadest areas for substantive cooperation between the United States and two countries with which the United States has had relatively cool relations. Would a change in the amount or type of assistance provided be beneficial to U.S. relations with Cambodia, Laos, or Vietnam? Should the U.S. government use UXO assistance to pressure other entities, such as Vietnam's MND, to be more cooperative in the UXO decontamination effort?
- UXO prevention—The Department of Defense has implemented a policy that is to eventually replace all cluster munitions with ones whose failure rate is below 1%. Should the U.S. government

undertake additional efforts to reduce the amount of post-conflict UXO from U.S. munitions, including prohibiting the use of U.S. funding for certain types of submunitions that may leave UXO? Given DOD's current views and policies on cluster munitions and landmines, does this preclude the United States from joining the 2010 Convention on Cluster Munitions or 1999 Ottawa Convention on Landmines?

- Precedents and lessons for other parts of the world—Are there lessons that can be drawn from U.S. assistance for UXO clearance and victim relief in Southeast Asia that may be applicable to programs elsewhere in the world, including Afghanistan and Iraq? Have the levels of assistance the United States has offered in Southeast Asia signaled a precedent for other parts of the world?

During the 115th Congress, legislation was introduced that would have addressed some of these general issues associated with UXO, though none directly addressed the current situation in Southeast Asia. The Unexploded Ordnance Removal Act (H.R. 5883) would have required the Secretary of Defense, in concurrence with the Secretary of State, to develop and implement a strategy for removing UXO from Iraq and Syria. The Cluster Munitions Civilian Protection Act of 2017 (H.R. 1975 and S. 897) would have prohibited the obligation or expenditure of U.S. funds for cluster munitions if, after arming, the unexploded ordnance rate for the submunitions was more than 1%.

INDEX

A

B

C

D

E

F

G

H

I

J

N

O

S

T

W

Z

Related Nova Publications

Key Congressional Reports on U.S. International Relations

Editor: James Makinson

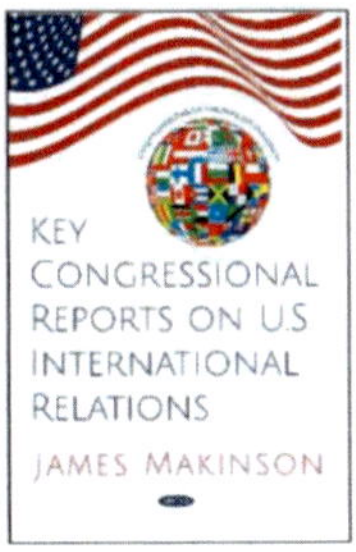

Series: Congressional Policies, Practices and Procedures

Book Description: This book is a comprehensive compilation of all reports, testimony, correspondence and other publications issued by the Congressional Research Service on U.S. International Relations during the month of March.

Hardcover ISBN: 978-1-53615-930-1
Retail Price: $230

Top 5 Reports and Testimonies in August 2018

Editors: Chad Johnston and Winifred Payne

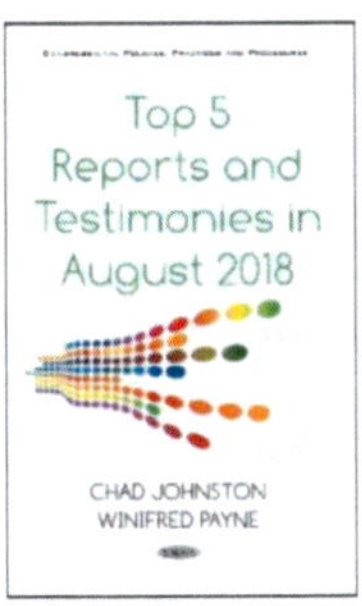

Series: Congressional Policies, Practices and Procedures

Book Description: This book is a compilation of significant reports and testimonies issued in August 2018.

Hardcover ISBN: 978-1-53615-511-2
Retail Price: $230

To see a complete list of Nova publications, please visit our website at www.novapublishers.com

Related Nova Publications

Key Congressional Reports for February 2019. Part II

Editor: Mandy Todd

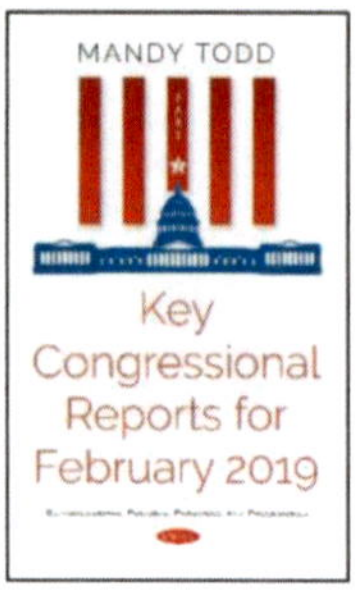

Series: Congressional Policies, Practices and Procedures

Book Description: This book is a comprehensive compilation of all reports, testimony, correspondence and other publications issued by the Congressional Research Service during the month of February, grouped according to topics. This book is focused on the following topics:
· Business
· Finance

Hardcover ISBN: 978-1-53615-737-6
Retail Price: $195

Key Congressional Reports on International Affairs

Editor: Hattie Ross

Series: Congressional Policies, Practices and Procedures

Book Description: This book is a comprehensive compilation of all reports, testimony, correspondence and other publications issued by the Congressional Research Service on U.S. International Relations during the month of February.

Hardcover ISBN: 978-1-53615-733-8
Retail Price: $230

To see a complete list of Nova publications, please visit our website at www.novapublishers.com